Provisional Bibliography of Atlases, Floras and Faunas of European Cities: 1600–2014

AF730020

John G. Kelcey

Provisional Bibliography of Atlases, Floras and Faunas of European Cities: 1600–2014

Springer

John G. Kelcey
Ceckovice
Bor u Tachova
Czech Republic

ISBN 978-3-319-31118-0 ISBN 978-3-319-31120-3 (eBook)
DOI 10.1007/978-3-319-31120-3

Library of Congress Control Number: 2016943118

© Springer International Publishing Switzerland 2016
This work is subject to copyright. All rights are reserved by the Publisher, whether the whole or part of the material is concerned, specifically the rights of translation, reprinting, reuse of illustrations, recitation, broadcasting, reproduction on microfilms or in any other physical way, and transmission or information storage and retrieval, electronic adaptation, computer software, or by similar or dissimilar methodology now known or hereafter developed.
The use of general descriptive names, registered names, trademarks, service marks, etc. in this publication does not imply, even in the absence of a specific statement, that such names are exempt from the relevant protective laws and regulations and therefore free for general use.
The publisher, the authors and the editors are safe to assume that the advice and information in this book are believed to be true and accurate at the date of publication. Neither the publisher nor the authors or the editors give a warranty, express or implied, with respect to the material contained herein or for any errors or omissions that may have been made.

Printed on acid-free paper

This Springer imprint is published by Springer Nature
The registered company is Springer International Publishing AG Switzerland

What may be my final publication
on urban ecosystems is dedicated to
Herbert Sukopp and Maciej Luniak.
My mentors and good friends
Of more than 40 years

About the Bibliography and Its Compiler

The bibliography contains 751 references of atlases, floras and faunas that have been published between 1600 AD and 2014 relating to 287 cities in what are currently the 47–56 countries of Europe. So far as is known it is the only bibliography of its type.

The bibliography provides information about the presence and distribution of plants and animals in cities throughout Europe. It will be of considerable interest to and should be used by a wide range of people including academics, researchers, librarians, school teachers, people with a general interest in the natural history of cities or just starting to be interested. The bibliography is an important tool for the professions involved in the planning, design and management of high quality urban developments, including biologists, architects, urban designers, planners, consultants, medics, sociologists, engineers, politicians, landscape architects, building surveyors, agronomists and landscape managers.

It is hoped that the bibliography will encourage people living in cities to start to take an interest in their surroundings and will cause those already interested to expand their interests to a wider variety of organisms and countries. Europe is a seriously neglected, little understood and disparate continent and yet it is the cradle of 'civilisation'. In short to break down the current insularity and nationalism that constrains the present lack of knowledge and thinking of natural history throughout Europe.

John G. Kelcey is not an academic but a practitioner of the art of managing natural resources. After graduating in Botany and Geology in the University of Manchester, Britain, he was appointed the ecologist and eventually also the Linear Parks Manager of the new city of Milton Keynes, the first such appointment in Britain. Subsequently he became an ecological consultant and advisor and carried out ecological assessments of major development projects mainly in urban areas, including roads, housing, commercial and retail developments and the provision of recreational facilities such as golf courses. He also prepared management plans, habitat creation schemes and supervised their implementation and has acted as a Resident Engineer. During this time he was a member of nine professional institutions and several learned societies throughout Europe. John G. Kelcey became interested in the natural history and geology of Europe in the early 1970s and has travelled extensively throughout Europe since that time. He has visited, worked in or lived in virtually all of the cities mentioned in the bibliography and many more besides to study their natural history and the arts and sciences of their countries.

Preface

This bibliography has had a very long gestation period; the original curiosity to know how much or how little information there is about the status and distribution of plants, animals and other organisms in urban areas throughout Europe was complemented sometime later by the development of three other major interests. First, a professional involvement in the planning, design and management of urban areas, second the colonization and evolution of hominids, other animals, plants and other organisms across Europe over the last 50,000 years and third the cultural development of hominids and their exploitation of natural resources. A fourth reason emerged from a wish to test whether there are sufficient comprehensive historical and current data to support statements about the biodiversity of cities. For example, in 2007, the Executive Secretary of the Convention on Biodiversity stated that '*the battle for life on earth will be won or lost in urban areas*.' The validity of this and similar statements made by others is entirely dependant on what is known about the presence/absence, distribution and relative abundance of organisms in cities. Initial research indicated that there are no comprehensive bibliographies/sources of information to answer these questions in relation to European cities.

For the reasons discussed in Chap. 1, there was one matter that was obvious from the outset, the bibliography would not be complete; the exercise would be one of using best endeavours.

Bor u Tachova, Czech Republic
June 2016

John G. Kelcey

Contents

1 **Prologue** 1
   Acknowledgements 1
   Scope of the Bibliography 2
   Importance of Floras, Faunas and Atlases 2
   Previous and Forthcoming Bibliographies 3
   Sources of Information 5
   Europe 5
   European Context 7
   National Context 7
   Biogeographical Regions 8
   Negative Information 8
   Period 8
   When Is a City an Urban Area? 8
   Extent of a City 9
   The 'City and District' v 'Around the City' 12
   Classification of Organisms 12
   Structure of the Bibliography 14
   Exclusions 15
   Geology 16
   Language 16
   Form of the References 17
   Conclusions 18

2 **Europe** 23

3 **Animals (Invertebrates)** 29

4 **Animals (Vertebrates)** 51

5 **Bacteria (and Related Organisms)** 79

6 **Fungi and Lichenised Fungi** 81

7 Plants (Non-vascular) 85

8 Plants (Vascular including Pteridophytes) 89

9 Protozoans and Viruses 113

Annex I 115
Individuals and Organisations Approached for References 115

Annex II 125
Number of Animals (Invertebrates) References for Each Country 125

Annex III 129
Number of Animals (Vertebrates) References for Each Country 129

Annex IV 131
Number of Non-vascular and Vascular Plants and Fungi References for Each Country 131

Annex V 133
Examples of General References 133

Index 135

Chapter 1
Prologue

It was evident from the outset that the preparation of this bibliography would raise many intractable issues. The solutions to the difficulties are matters of judgment, which will be resolved differently by different people, giving rise to endless and rather fruitless arguments and discussions. In short there is no right or wrong approach. No doubt my detractors will draw attention to the inevitable inconsistencies, such as why is this reference included and this reference not. All I can do is to say 'Here you are, this is what I have done and why I have done it, however inadequate and unsatisfactory it is'. The alternative is ignorance.

Acknowledgements

I am philosophically disposed to the view that those who have been especially helpful to an author or an editor should be acknowledged at the earliest opportunity and not as 'footnote' in the last paragraph where most readers never get to (ask a newspaper editor). And so it is that I wish to express my eternal appreciation of the considerable help that the following have given to me. As always, it has to start with three people, first my partner Liz who warned me about the task ahead should Springer graciously agree to publish the bibliography. Without her unstinting help, denial of sunbathing and other enjoyable activities the task would never have ended. Joint second are Herbert Sukopp and Maciej Luniak to whom I have dedicated this publication but they deserve much more from me. And then there are 40 others who have been of invaluable and exceptional assistance in supplying references, translations, photographs and other information, I shall be eternally grateful to them. Sadly, there are too many to list individually so I give them my apology and hope they understand my dilemma, if not embarrassment.

© Springer International Publishing Switzerland 2016

J.G. Kelcey, *Provisional Bibliography of Atlases, Floras and Faunas of European Cities: 1600–2014*, DOI 10.1007/978-3-319-31120-3_1

Scope of the Bibliography

This bibliography covers atlases, floras and faunas (the animal equivalent of floras) that relate to a whole 'city' (as defined below). As described later, accounts of plants or animals of a part of a city or of a particular 'group' of organisms are not included. To avoid frequent repetition of 'atlases, floras and fauna' after the following section 'atlas' is used as the collective noun for all three.

Importance of Floras, Faunas and Atlases

The early accounts of plants were mainly 'herbals,' whilst the early faunas were mainly concerned with pest species and those of economic value. During the nineteenth century floras and faunas were produced as the result of an increasing academic and general interest in natural history. Since the 1950s most atlases, floras and faunas have been prepared for nature conservation reasons. However they have a much more important role in the management of natural resources in cities, including human welfare. Many organisms are of considerable benefit or interest to people and therefore considered to be desirable. On the other hand a large number are directly or indirectly detrimental to human interests and welfare and therefore considered to be undesirable to people although they may be important to biological systems as a whole. In both cases it is essential to know where they are, their status and dynamics.

As a general rule the most frequently produced atlases, floras and faunas are mainly concerned with the 'desirable species', which include the larger, more colourful and therefore conspicuous organisms such as the macro-lepidoptera, amphibians and reptiles, birds, terrestrial mammals and vascular plants. These 'groups' comprise species that are easy or relatively easy to identify and are of interest to many people; consequently there is a relatively large 'recording/surveying pool'. The undesirable species are recorded by default.

The major omissions concern the 'groups' of inconspicuous organisms that are desirable and undesirable or contain species in both categories. The first group includes the Orthoptera (crickets and grasshoppers) and the Annelida (worms); the undesirable groups include Acari (mites and ticks), Arachnida (spiders), Coleoptera (beetles) and Siphonaptera (fleas). Those groups that contain desirable and undesirable species include the Diptera (flies), the Hymenoptera (bees, wasps and ants), the fungi and the algae. With the exception of the macro-fungi, most of the organisms are small (often microscopic) and not or rarely seen. Most of the groups contain a very large number of species, most of which are difficult to identify, and the number of people able to identify them is very small. The importance of these organisms is disregarded by politicians and public administrators probably because they are inconspicuous and consequently their importance is under-estimated, not appreciated or understood. These are very serious failings.

It is unfortunate that public administrators (including some scientific administrators) do not see the need to anticipate biological problems preferring to react to them. Atlases allow them to be prepared and to plan—whether it is in relation to the decline of or increase in a species for human, plant or animal health. For example, *Rattus* spp., which carry fleas (Siphonaptera), which are the vectors of the bacterium *Yersinia pestis*—the cause of bubonic plague that devastated the human population in Europe. In recent times the fungi *Ophiostoma ulmi* and *O. novo-ulmi*, which are carried by the bark beetle *Scolytus* spp. destroyed the Elm flora (*Ulmus* spp.) of parts of Europe. The list is 'endless' but the fact of the matter is that 'we' have very little information about the distribution of organisms in the cities of Europe, which means that they cannot be managed effectively in the long-term interests of their inhabitants.

It is essential to issue a 'health warning' in relation to the interpretation of 'dot' maps— their value and importance is in showing the distribution of a taxon and not its abundance. Many biologists and non-biologists interpret one dot in say 1 % of a 10 km × 10 km grid map of a city as meaning the taxon is very rare whilst a dot in say 90 % of the squares means that a taxon is abundant. Put another way a dot represents the occurrence of a taxon not its abundance, consequently the 'dot' may represent 1 or 100,000 individuals. The difficulty is exacerbated by changes in scale whether a decrease from say 10 km × 10 km to 2 km × 2 km to 1 km × 1 km to 100 m × 100 m or the converse. The higher the scale the higher the over-recording, often by several thousands of percent.

Previous and Forthcoming Bibliographies

In 2002 Clemants published *A Short Bibliography of Urban Floras* (Urban Habitats 1(1) 25–29). The bibliography, which is on a worldwide scale and restricted to vascular plants, is extremely short; it lists 16 references relating to 13 European cities in 11 countries; however not all of them relate to the whole city. The author's plea, '*I welcome any additions or corrections to this list*', suggests that he was aware that the bibliography was not comprehensive.

As far as can be ascertained Clemants is the only bibliographical reference concerned with floras. In a recent paper *'Pioniere der Adventivfloristik: Einige Skizzen als Anregung zu einem biographisch-bibliographischen Projekt' (Pioneers of the Adventive Flora: Some sketches are inspiration for a biographical and bibliographical project) (Braunschweiger Geobotanische Arbeiten 11: 155–171)* Sukopp and Kasperek have indicated the aspiration to produce a comprehensive bibliography on the geography and ecology of alien plant species. So far as animals are concerned a review of atlases (worldwide) of terrestrial birds was published by Dunn and Weston in 2008—(Emu 108: 42–67). An account of the ornithological atlases in Europe by Maciej Luniak is to be published in Murgi E and Hedblom M (eds) *Ecology and Conservation of Birds in Urban Environments. Springer (in press).*

Several bibliographies have been published in relation to the more general aspects of urban ecology. In 1990 the Centre for Urban Ecology in Birmingham, England, listed 1768 worldwide references (reported to be a representative sample) on urban ecology that were published between 1980 and 1987. Herbert Sukopp and his colleagues found a total of 6500 references (mainly in German) relating to urban ecology *sensu lato* listed in several bibliographies published between 1986 and 2000. In 2008 Werner and Zahner published an excellent review and bibliography under the title *Biological Diversity and Cities—a review and bibliography*, which lists 648 publications that were published throughout the world between 2000 and 2008. They mention that in 1988 alone more than 600 papers were published on the urban avifauna of Italy.

An informed guess based on this information, general experience over the last 40+ years, the publications in my own library and the references listed in them suggests that between about 1600 AD and 2014 of the order of tens (maybe extending into six figures) of thousands of books and papers on the natural history (*sensu stricto*) of cities have been published worldwide, excluding 'local' publications and individual contributions to books. However it seems that only a very small proportion (may be less than 1 %) are likely to be atlases.

If books and papers on the urban environment *sensu lato*, for example, engineering, architecture, micro-biology, medical, air and water pollution, geology, soils, finance, planning and social development (and their subdivisions) are added, it is likely that number would be trebled, if not quadrupled. This represents an enormous amount of information but are the professions involved in the urban environment any the wiser in terms of the cohesive management of natural resources in the planning, design and management of cities?

A major difficulty is that the information is largely fragmented in time and space; consequently it is not easy to obtain. As a result it is of little use in the understanding of the structure and function of urban ecosystems and consequently their planning, design and management. Because of the often significant differences in approaches between European countries it is often difficult, if not impossible, to extrapolate from one to another. A botanist or zoologist brought up and educated in one country is likely to have a different perspective than one educated in another. Common sense suggests that this cultural aspect is likely to have a strong influence on the preparation of atlases. The difficulty is exacerbated by some differences of approach within some countries. As a consequence the data may not be comparable.

Information about the factors controlling the distribution, frequency and other biological aspects of plants, animals and other organisms is generally based on national studies, which is a rather insular approach. Taxonomy is a classic example of the principle; as considered later, taxonomic studies of organisms need to be carried out on a European scale and often beyond. A similar principle applies to the preparation of atlases.

Sources of Information

The exercise was going to be highly dependent on many people throughout Europe; consequently I contacted all those organisations and individuals I knew who had or might have knowledge of the urban ecology of Europe. In addition I approached individuals and organisations I did not know to ask if they could assist or knew anyone who might be able to do so. The 200+ organisations and individuals contacted are listed in Annex I, my apologies for any omissions. Many of those contacted were helpful in providing information or directing me to others and I wish to thank them very much for their help. On the other hand quite a lot, especially public organisations such as museums, did not reply. Some light was shed on the reason in relation to some countries of central and eastern Europe during a discussion with the cultural attaché of an eastern European country about a similar difficulty concerned with my research of contemporary visual arts. 'Why', she said in somewhat severe tones, 'Do you expect anyone from my country to reply to letters—they simply do not; it has nothing to do with language, it is cultural'. Several months later she reiterated the same point but in advisory terms, 'Please do not expect people from my country to reply to letters, they do not even reply to me'. In short, there is another cultural issue.

Language also plays an important part; because of my British education my knowledge of modern European languages is minimal and it is wrong of me to seek information in English when many of the recipients of requests have no or at best little knowledge of it.

Several people have supplied references that have not been incorporated because they are not appropriate to this bibliography. I shall be eternally grateful for their help and the considerable effort they have made. There is no doubt that they will not be pleased; I take full responsibility for their disappointment and annoyance; I can only hope that if they read this Prologue they will understand the reasons, although they may well disagree with I am very pleased that the proposal by Sukopp and Kasperek has come to my rescue—see the section 'Sources of Information' above.

A large number of publications including books, papers and annual reports of natural history societies (in Britain) have been examined for references to atlases. These publications are not listed because the number of pages needed would substantially exceed the number of pages of the bibliography itself.

Europe

During the slightly more than four centuries covered by this bibliography the definition of Europe and the number of countries in it has changed dramatically, especially in the few years following 1918 and again shortly after 1945. In more recent times further divisions have occurred, for example the break-up of the former Soviet Union, Jugoslavia and the division of Czechoslovakia. Several countries are on the verge of dividing into independent states; this gives rise to an inconsistency of the definition of a ‘country’ for the purposes of this bibliography; for example, Britain is now virtually divided into four countries and is treated in this way, whilst Spain (which is close to being divided into four countries) is treated as a single country although it could be argued that it should also be considered as four independent political units.

As a consequence of political dynamics, Europe has become a highly complex continent. For some purposes Europe is considered to include one or a combination of Greenland, Israel, Kazakhstan, Turkey and ‘North Africa’. On the other hand the definition of Europe is sometimes restricted to the European peninsula with or without Scandinavia. The consequence of these different approaches is that Europe may comprise between 47 and 56 countries. The lower number excludes Armenia, Azerbaijan, Georgia, Greenland, Israel, Kazakhstan and Turkey because they are not always considered to be part of Europe and Kosovo and Vatican City, which are not recognised as independent countries by some countries.

Defining the southern boundary of Europe west of Greece is not without problems. Common sense suggests that the Europe/Africa boundary should be a line midway between the coastal areas of the two continents; however it could be the low-tide mark of the two continents or 12 miles from the low-tide mark, leaving the area in between as international waters. The complications do not end there because the Canary Islands, which are politically a part of Spain (but outside the European Union), are at a lower latitude than the southern margin of the Mediterranean and more related to Morocco in biogeographical terms. Cyprus is a unique problem because it is governed by one state that is in Europe (Greece) and one that is not (Turkey); to make matters worse (in terms of Europe) the island is nearer to Turkey and Syria than it is to Greece.

Fortunately these difficulties do not arise because there appears to be no urban atlases for cities on these islands. Nevertheless there seems to be a case for having two definitions of Europe, a political one and a natural one—it is a principle that can be applied to other natural/political-administrative issues. No doubt there would be very strong objections to considering Cyprus as part of Syria or Turkey and the Canary Islands as part of Africa even if the division was made for biological purposes only.

The most commonly used definition of Europe and the one that is used in this bibliography is:

> the area between the Atlantic Ocean and the Ural Mountains and between the Mediterranean and the Arctic Circle.

This definition results in 52 countries (listed in Annexes II–IV), of which more or less half are members of the European Union.

European Context

In order to provide a European framework against which the 'atlases' of individual cities can be assessed the bibliography starts with references to the geology, geomorphology, major habitats and atlases relating to Europe as a whole. However the two sets of data are not directly comparable and therefore they can only act as a general guide.

Because of the lack of recent 'atlases' or because those that have been published are in multi-volumes and are therefore expensive Collins Field Guides of Britain and Europe have been quoted whilst those related to a part of Europe for example 'Northern Europe', have been excluded. British field guides have been used for reasons of language and ease of access. Similar and sometimes more preferable field guides have been published in most of the other countries and can be substituted.

National Context

National 'atlases' and information about the geology, geomorphology and soils to provide the context for assessing the individual cities in relation to the countries they are in have been excluded for several reasons. First, the bibliography is concerned with Europe as a single ecological/geographical unit. Second, the substantial changes to the countries of Europe and their extent during the 414 years covered by the bibliography make the provision of such information impractical. Third, these changes complicate the 'European picture' and are therefore not helpful. Fourth, the difficulties involved in obtaining references. Fifth, the provision of such references would increase the length of the bibliography substantially for little, if any, advantage.

Another complication that has fortunately not been encountered during the preparation of the bibliography (although it may affect future bibliographies) is the name changes of cities, for example, Bratislava, which has been known by five different names (two are still in general use). In some countries older names are being reintroduced for example, Sibiu=Hermannstadt (Romania), Aachen=Aix en Chappelle (Germany) and Liege=Luttich=Luik (Belgium). In some cases the name of a city has changed as a consequence of internal political decisions; for example St. Petersburg became Leningrad and has since been renamed St. Petersburg—however caution is needed because 'Leningrad' is still used as the name of the region. Other name changes have resulted from changes in national boundaries; for example Gdansk (Poland) was formerly Danzig (Germany) whilst Beograd (Belgrade) was once in Jugoslavia but is now in Serbia.

These issues have been resolved by converting the headings to the name of the city that is in general use. In countries where the Cyrillic alphabet is used, the Roman equivalent is given. In cases where one or both continue to be unclear, the current English name is used—as a last resort.

Biogeographical Regions

Given the dynamics of the geopolitical boundaries of Europe, the inevitable future changes and the fact that organisms are not constrained by such boundaries there is considerable scientific merit in using the biogeographical regions as standard units in relation to studies of the distribution of plants, animals and other organisms. This applies even to the extent of converting data from previous decades, if not centuries. The simple point is that the earth and biological sciences do not mix with geopolitics and nationalism.

Negative Information

The bibliography contains a substantial amount of negative information. Contrary to general scientific thinking negative information is extremely valuable in making people aware of what is not known—in the case of this bibliography it is a lot in terms of cities and organisms. The use of the term 'not known' is used to bring the message home. There is also the sad psychological point that many people only react to 'not known' to demonstrate that the author(s) is incorrect or lacking in knowledge. Although it was not tried in this bibliography, the author has found that one of the better ways of obtaining information is to state that it is not available—then follows a flood of it.

Because of the large number of Classes and Orders of invertebrates negative information is not included. The reader should therefore assume that if a particular class or order is not mentioned then an 'atlas' reference(s) has not been found, not that one does not exist.

Period

The end period (that is, 2014) was easily determined. The more difficult task was to determine the start. Initially 1800 was considered to be the more appropriate date but preliminary investigations indicated that there are a few important earlier references; consequently c.1600 was chosen as the start date.

When Is a City an Urban Area?

In this bibliography 'city' is used as a synonym for an urban area of any size because it is shorter and less 'ugly' than the alternatives. However it is probably better in the first instance to consider the question of 'when is an urban area an urban area' and therefore 'when is a city, a city?' Human settlements comprise a series of overlapping continua—the ephemeral to the permanent; a single building to millions of buildings; 100 m^2 of hard surface to hundreds of square kilometres; a population of say 25 people to more than 25 million and growing. The human brain appears to be unable to cope with continua and consequently 'things' have to be classified. This gives rise, as do all classifications, to endless and rather pointless arguments as to where to draw the line. Numerous attempts have been made to define a 'city' but they are all arbitrary, matters of judgment and therefore irrelevant. Everyone knows when a 'city' is a 'city' (or an 'urban area' is an 'urban area') just as everyone knows when a house is small or large, without the need for quantitative data. The definition of a city also varies between disciplines, for example archaeologists use 'city' to describe developments that are substantially smaller than the definitions more commonly used by ecologists.

Extent of a City

There are two aspects to assessing the extent of a city: first, the administrative boundary and second, the limit of the built up area, (Fig. 1.1). In most (if not all) cases there are environmental and other differences between the edge of the urban

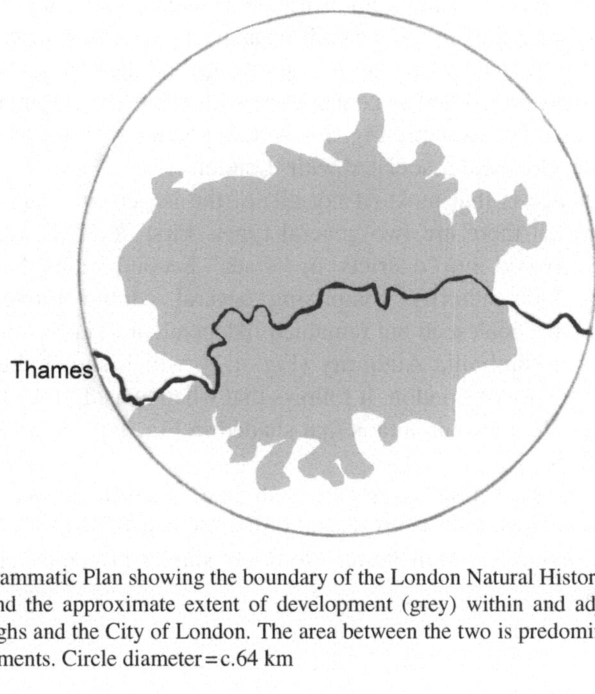

Fig. 1.1 Diagrammatic Plan showing the boundary of the London Natural History Society survey area (circle) and the approximate extent of development (grey) within and adjacent to the 32 London Boroughs and the City of London. The area between the two is predominantly rural with scattered settlements. Circle diameter = c.64 km

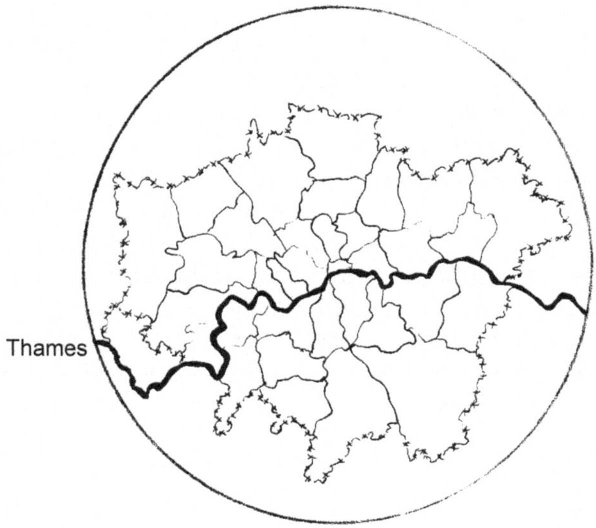

Fig. 1.2 Diagrammatic Plan showing the recording area of the London Natural History Society survey area (circle) and the extent of the Greater London Authority (x-x-x), which includes the 32 London Boroughs and the City of London). Circle diameter = c.64 km

development and the political/administrative boundary. In most cases urban development does not extend to the administrative boundary resulting in a fringe of 'countryside' between the limit of the built-up area and the administrative boundary. In other cases the urban development has spread beyond the city boundary into the adjacent rural areas or resulted in coalescence with urban development in another administrative area. For example Fig. 1.1 is a diagrammatic representation of the edge of urban development associated with 'London'.

The consequence is that most (if not all) of the larger cities are divided into 'subunits' of which there are two general types. First, a single administrative/political unit is divided into 'districts' or 'zones'. Second, cities that comprise a strategic 'umbrella' authority comprising several administrative/geopolitical authorities that have coalesced but remained independent of each other and semi-independent of the umbrella Authority (Fig. 1.2), which shows the 32 London Boroughs and the City of London. It follows that when considering issues relating to urban ecology the relevant area is that shaded in Fig. 1.1 and not the circle or the GLA boundary.

In summary what constitutes a city varies in time. The difficulties in determining what to include and exclude from the bibliography can be demonstrated by two British cities, London and Manchester. No doubt similar principles apply to other European cities.

As a consequence of its rapid expansion during the nineteenth century the London County Council was created in 1888 and remained in existence until 1964 when it was replaced by the Greater London Council, which comprised 33 geopolitical units (32 'London Boroughs' and the City of London), see Fig. 1.2. The Greater London Council was abolished in early 1986 and replaced in 2000 by a strategic umbrella organisation called the Greater London Authority.

The more recent geopolitical changes in Manchester present a much more intractable difficulty in relation to the compilation of this bibliography. Prior to 1974 the 'City of Manchester' was an independent geopolitical entity that was contiguous with or close to eight other independent geopolitical entities, which had been formed from the merging of small cities during earlier times. In 1974 the Government created a single entity (the 'Greater Manchester City Council', which comprised ten 'sub units'—the eight authorities previously referred to and the creation of two other autonomous authorities that did not previously exist. However in 1986 a subsequent Government decided to abolish the Greater Manchester Council and create ten independent geopolitical authorities—eight for the 'second' time and two for the first. In 2011, the Greater Manchester Authority was established as a strategic organisation covering ten geopolitical units that are semi-independent of it—a similar situation to that of London. In biological recording terms the 'Manchester situation' can be summarised as:

1. Before 1974—eight cities
2. 1974–1986—one city
3. 1986–2011—ten cities
4. 2011 to present—ten with an 'umbrella' strategic authority

The two examples raise the issue of what should or should not be included in the bibliography. It is reasonable to argue that despite the 13-year gap 'London' has, *de facto*, been a single 'biological' entity since 1888. On the other hand Greater Manchester is of relatively recent origin having been created in 1974; it is reasonable to ignore the period of 25 years when it did not exist. Consequently references relating to what are now the components of Greater London before 1888 and Greater Manchester before 1974 have been included whilst references after those years have been excluded. A word of caution is needed because 'Greater' is often omitted, as it is for other cities elsewhere in Europe.

Usually the biological recording unit extends to the edge of the administrative boundary but in some cases the recording unit is an arbitrary distance from the centre of the city. For example the London Natural History Society's recording unit is 20 miles (32.2 km) radius from St. Paul's Cathedral, which extends well into the surrounding countryside and includes many separate urban settlements. The same principle applies to other areas such as the Sorby Natural History Society in Yorkshire.

The difficulties in interpreting the information could and should be overcome by demarcating the limits of development, the administrative boundary and the boundary of the recording unit.

The 'City and District' v 'Around the City'

Some publications have titles such as the 'Plants of City X and District'. This implies that the atlas contains an account of the plants, animals or other organisms in and adjacent to the city. Other publications have titles such as the 'Plants Around City Y', which suggests that the publication is concerned with the area around the city NOT within it. In both cases it is likely that all or at least part of the 'district' or 'area around' will now be part of the city; the older the reference the greater the probability of this being the case. Notwithstanding this situation, publications with the title 'Plants of City X and District' have been included whilst the latter have not.

Some publications cover a large number of cities scattered throughout a matrix of countryside, for example, Travis' Flora of South Lancashire (part of north-west England), which was published in 1963. Since that time there has been substantial urban development including the expansion of existing cities and the creation of new ones, some of which have coalesced, for example the components of what is now Greater Manchester. Although the southern part of the area was and remains substantially urban the ratio of rural to urban land is too great to include the Flora in this bibliography.

As considered earlier in relation to St. Petersburg (the city) versus Leningrad (once the city and region and now the region only) there are many cases where great care should be taken when considering the relevance of a reference to a particular city, for example the Canton of Geneva is ×20 the size of the city of Geneva. Several investigators appear to have relied on references that suffer from these and some of the other difficulties considered above. For example, they have used atlases (*sensu stricto*) of predominantly rural areas that include cities to examine botanical issues in urban areas. A similar approach has been taken by some ornithologists.

Where a city has been a single unit, then two political units which have been reunited sometime later, publications of the single and two units have been included. Berlin may be a unique example.

Classification of Organisms

Because the bibliography has a biological or more precisely a taxonomic structure it is desirable to provide a simplified explanation of systematics and taxonomy (the classification of organisms), for users who may not be familiar with them. The classification is a hierarchical sequence (rank) which in general terms starts with 'kingdom' and ends with 'form'. However some classifications have a rank (domain) above kingdom:

Kingdom
Phylum
Class

Order
Family
Genus
Species
Variety
Form

The ranks between and including phylum and family may be qualified by 'super/supra' or sub-; the latter is also applied to 'species'. The ranks and individual organisms are 'taxonomic units', which are more commonly known as 'taxa' (singular—taxon). The name given to and the placing of an organism in a particular rank is a matter of the judgment of a taxonomist (the name given to biologists who specialise in the classification of organisms).

Even at the highest rank there are considerable arguments—at the time the bibliography was completed there are between five and seven kingdoms (excluding viruses, which is not a kingdom—yet!). There are also several different sequences for placing an organism in a particular rank and its name may vary from country to country or even within countries. The normal scientific sequence is taxonomic and related to evolutionary processes for instance fish, amphibians, reptiles and mammals.

This sequence is easy but it becomes increasingly more difficult with invertebrates and plants; for example a botanist specialising in vascular plants may not know the scientific sequence of the invertebrate orders, equally an invertebrate zoologist may be unfamiliar with the sequences used in non-vascular and/or vascular plants. Because it is easy for everyone the bibliography uses the A–Z sequence of the Roman alphabet, which most people understand, but it is likely to enrage many biologists, to whom the taxonomic sequences are sacrosanct.

'In recent times there have been some changes or proposed changes to the classification of plants, as follows:

1. Marine algae, the changes do not affect the bibliography.
2. Dinophyta (Dinoflagellates) are now protozoans.
3. Cyanophyta (blue-green algae) are now bacteria.
4. Bryophyta have been separated into three phyla:
    (a) Bryophyta—mosses
    (b) Marchantiophyta—liverworts
    (c) Anthocerophyta—hornworts
5. Vascular plants (the vernacular term used in this bibliography) are divided into
    (a) Lycophytes (club mosses and quillworts)
    (b) Euphyllophytes, which are divided into
        (i) Monilophytes (ferns and horsetails)
        (ii) Spermatophytes (Conifers and flowering plants)

Structure of the Bibliography

Four main ways of constructing the bibliography were considered, two geographical, one biological and one that comprises both. The first option was to list the cities in an A–Z sequence independently of the countries in which they occur and to list the references in taxonomical (or in an A–Z) sequence: Cities-Taxonomic. The second option was to list the references in the sequence Country-Cities-Taxonomic. The third option was to start with the taxonomic status followed by the country and then the city, Taxonomic-Country-Cities. A fourth option would have been to produce the bibliography in two parts, Part I comprising option 2 and Part II comprising option 3.

Although option 4 is the better one it was rejected for reasons of cost and logistics because it would require the preparation of what in essence would be two books. The first option is unsatisfactory mainly because it does not relate the city to a country and was therefore rejected. This left a choice between options 2 and 3; choosing between them was based largely on considering the approach that the majority of the potential users would prefer. Both options have the disadvantage of being too narrow; option 2 is nationalistic whilst option 3 promotes a 'specialist' approach.

After much consideration option 3 was adopted on the basis that most biologists are primarily interested in a particular 'group' of organisms in many countries rather than many groups of organisms in a particular country. This approach surprised the publisher who considered the converse would be the case and therefore preferred the geographical approach of option 2.

The initial division of the bibliography is into kingdoms (animals, bacteria, fungi and plants), protozoans and viruses. Using animals as an example, the animal kingdom is divided into invertebrates and vertebrates, which are further subdivided by class or order (in A–Z sequence); for example the vertebrates (a sub-phylum) are divided into five classes—amphibians, birds, fish, mammals and reptiles. The countries are listed under each class or order followed by the city to which the reference relates. The references are listed in chronological order. A worked example:

- Animals
  - Vertebrates
    - Birds
      - Germany
        - Berlin
          - References

Exclusions

The primary objective was to prepare a bibliography of published atlases of classes or orders relating to the whole of a city, in summary 'macro-scale' and 'macro taxa', rather than to parts of cities and/or the lower ranks of taxa. This raised many difficult questions about what references to include and what to exclude but the line has to be drawn somewhere in order to sustain the objective. The answers resulted in some arbitrary decisions. For example, references that refer to 'central area' have been excluded as have references relating to 'groups' such as raptors, ducks and plants such as ruderal or synanthropic species. The same principle applies to publications relating to a particular species or a 'group' of species that occur in a specific habitat. Because it is normal practice to produce atlases of breeding and over-wintering birds separately both are included. However, although atlases of terrestrial and aerial (bats) mammals are generally published separately and after a lot of thought an arbitrary and some will argue an illogical decision was made to include only those publications that refer to 'Mammals'.

A similar but much more intractable problem arises with most invertebrates, where the 'tradition' is to publish 'atlases' of orders, suborders and even families. Vertebrate zoologists and botanists may seek to argue that 'what is good for the invertebrates is good for vertebrates and plants' but given the number of Classes, Orders and Families in the invertebrate fauna it would not be reasonable or practical to produce 'atlases' for taxonomic ranks below the rank of Order'. However, there is one exception to this general approach, namely the inclusion of references relating to families where they form part of a series, for example Warsaw.

Many of those approached for information made strenuous requests for the inclusion of references relating to synanthropic plant species. There is no doubt that they will be bitterly disappointed, if not annoyed, that I have not done so, the main reason being that it would create a precedent in relation to other 'subdivisions' of plants, animals and other organisms and substantially extend the self-imposed objective, the implications of which would be to increase the time and effort involved in the preparation of the bibliography and the cost of doing so. Hopefully their disappointment will be compensated for by the proposal outlined in the paper by Sukopp and Kasperek, referred to earlier. The major difficulty in including atlases relating to part of a city, synanthropic plants and other 'subcategories is that it undermines the fundamental objective of the bibliography and would open 'the flood gates' and probably increase the effort involved exponentially.

Unpublished written/typed manuscripts have been excluded as have atlases, floras and faunas that have only been published on websites or referred to as 'in press or in preparation'. Dissertations have also been excluded because they are generally difficult for people to obtain—there is a good case for some of them to be published by the university so that they may be immediately available to interested people. References with insufficient information have also been omitted.

Geology

A major omission is the exclusion of accounts/'atlases' of the geology and soils of cities. There are two elements to the subject, first the distribution of natural rocks, minerals and fossils and second building stones.

Many cities are built on flat or relatively flat land with few if any geological exposures, for example Rotterdam and London. Nevertheless accounts of the geology of some such cities have been published, for example, Whittaker W (1889) The Geology of London and AAVv (1995) La geologia di Roma. Mem. Descr. Del-la Carta geologica d'Italia.

All cities have one 'geological formation' in common, namely building stones of which there is a large variety including sedimentary, volcanic and metamorphic that vary from acid to alkaline. The rock type determines some of the plant species that are able to colonise and establish on buildings and other structures and therefore some of the animal species present. Whilst there are a few short accounts of the building stones that occur in some cities (for example Simpson and Broadhurst's 'A Building Stone Guide to Central Manchester', published by the University in 1975) there appears to be no comprehensive atlases showing the distribution of the rock types used and their distribution in any European city. Geology is an extremely important aspect of the natural history of cities and fascinating and instructive in its own right but sadly it appears not to attract much interest from biologists and people at large.

The literature is neither comprehensive nor extensive and yet mapping the distribution of building stones is probably much easier than preparing atlases for organisms. Metamorphic, volcanic and sedimentary rocks have been used for the construction of many different types of building. In the past limestones were often used for prestigious public buildings; for example Tinnye Limestone was used to build the Parliament buildings in Budapest. In more recent times volcanic and metamorphic rocks have been used quite commonly, for example Serpentinite and the light and dark forms of Larvikite.

Mapping the distribution of natural deposits and building stones can be extended to the preparation of atlases of fossils and trace fossils in cities.

Language

The compiler's lack of fluency of the various alphabets and languages used throughout Europe has been a significant constraint on the finding of references and determining their relevance to the bibliography. The result has been a regrettable and embarrassing bias towards publications with English titles or which include English titles. The language constraint has many implications for the preparation of the bibliography, including whether a publication or a quoted reference should be included or excluded on the basis of the criteria discussed above.

In some cases it is not known whether the title of a reference quoted in a reference list is in the original published language or an English translation. In other cases whilst the title is in English in the reference list the original publication may be in another language. These difficulties are exacerbated in those countries that use the Cyrillic alphabet not least because these days Greek and Latin are rarely taught in schools whilst very low priority is given to the teaching of languages; it appears that linguistics is not taught at all.

In order to overcome the difficulties described in the previous paragraph, the approach adopted in this bibliography is to give the reference in the original language (as far as it is known) with the title being given in brackets in English. However in some cases where a paper is published in a country that uses a form of the Cyrillic alphabet only the English translation of the whole reference may be available.

Form of the References

Where possible references were taken from the original publication but in most cases they are quoted as they appear in the publications from which they were extracted. Many books are reissued as reprints, revisions and impressions or in two or more editions. The dates of reprints, revisions and impressions are not given whilst the inclusion of the edition number is inconsistent; sometimes it is and sometimes it is not. During the preparation of the bibliography it became apparent that what turned out to be the same publication was quoted differently in as many as three to four reference lists. In some cases the author was different or the sequence of authors varied as did the year of publication and most disturbing there were substantial differences in many of the titles. The latter caused major problems because the published title indicated that the reference should be included in the bibliography; however when it was checked against another list or with the original it was discovered that the reference was inappropriate. For example, the subtitle qualified the main title for example, by stating that it related to the 'area around' but outside the city. In some cases the year of publication and the publisher are different or not known. There are inherent inconsistencies caused by different authors, editors or publishers using different formats (including punctuation), use of hyphens and variations in the use of upper and lower case letters.

Translations and the lack of knowledge may also have caused some inconsistencies. For example, a reference to Bonn did not include the city, whilst references to Leningrad that were thought to apply to what is now the city of St. Petersburg actually relate to the Leningrad Region not the city. An example of this difficulty has been left in the bibliography to demonstrate the point. Two references are given under Orthoptera (Germany) – one is for Hessen and the other for Giessen. When compiling the bibliography it was thought that Hessen was a city; it was only when correcting the proofs that a German colleague indicated that Giessen is a city within the Region of Hessen. It is therefore suggested that references to a city should be followed by the region in brackets, in this case Giessen (Hessen).

It follows that copying references from publications and not taking them from the original publication are likely to have resulted in continuing most of the errors identified in the previous paragraph. Where the same publication is given differently in two and often more different reference lists and where it has not been possible to check the original the reference included in this bibliography was chosen more or less arbitrarily. As a consequence there will be significant differences between some of the references quoted in this bibliography and those quoted in a reference list elsewhere.

Conclusions

Although accounts of the fauna and flora of cities occur in the literature of ancient Greece and Rome they appear to be relatively rare. There seems to be a virtual absence between the middle of the first millennium AD and the end of the sixteenth century. It could be that more are waiting to be discovered but it will require expertise in biology and languages (ancient to modern), which is extremely rare.

Floras, in the form of 'herbals', started to appear in the sixteenth century. The increase of interest in natural history in the subsequent centuries resulted in a concomitant increase in the publication of floras and faunas, generally related to the medical sciences. As time passed medical science and natural history diverged (which was and remains a tragedy of serious proportions). A growing interest in literature and science in the nineteenth century resulted in the creation (in Britain, at least) of Literary and Philosophical Societies (with a natural history subdivision) and natural history societies. The curiosity and interest in plants and animals stimulated these societies to prepare and publish accounts of the floras and faunas of the cities in which they were founded.

By the mid-twentieth century a new approach was being developed, commonly known as 'dot maps', where a dot represented the occurrence of a species within a rectangular grid. The 'atlas' was born. However floras and faunas (with and without 'dot maps') continued and continue to be produced although in some relatively recent cases a complex variety of symbols have replaced the dots.

Although the number of countries that comprise Europe has varied over the last 414 years, the continent currently comprises 47–56; this bibliography uses 52. The size varies from small islands such as Malta to large land masses such as Russia (whether before or after 1989). A recently published road atlas lists 7000 cities throughout Europe, which common sense suggests is a significant under-recording. This is confirmed by an assessment of the 1:1,000,000 scale Michelin map of France published in 2013, which shows and lists 3070 cities and if extrapolated to the whole of Europe and adjusted for the different sizes indicates that the number of cities in Europe is likely to be of the order of 50,000, if not more. Whilst both statistics are unreliable in themselves they indicate, even allowing for an overestimate of the number of cities and the omission of a substantial number of references, a depressing picture namely the serious lack of atlases of the distribution of organisms in European cities.

A superficial assessment of the bibliography indicates (allowing for its deficiencies) that only a very small number of atlases have been published over the last 414 years or thereabouts: a smaller fraction are extant and of those a tiny percentage are 'recent' (that is published in the last 10 years). There is a marked bias in the organisms (most of the atlases are concerned with birds or plants) and countries, for example most of the atlases published in Italy are concerned with birds.

There seems to be two major causes for the paucity of atlases—the lack of funds and with a few exceptions the lack of expertise. The preparation of most, if not all, atlases relies on unpaid botanists and zoologists to collect the data—often and regrettably generally called 'amateurs'. It is an offensive and pejorative word that is demeaning to describe a large number of people who give freely of their time and expertise, which is often greater than that of some 'professionals'.

Atlases are by reason of their preparation historic documents that are out of date before they are published; it may take up to 10 years (sometimes longer) from conception to publication. Revision intervals vary from one to several decades. The revisions themselves may take up to 10 years depending on the lapse of time since the previous atlas. Even with sufficient resources it is likely to take a minimum of 3–5 years to produce an atlas because of the logistics of obtaining the data (access to land, water and buildings, availability of sufficient expertise and the need for licences in relation to protected species) allowing for the seasons and annual variations followed by the analysis, coping with taxonomic changes and the preparation of the data and finally publication. This does not demean the value of atlases providing that they are not used as an indication of the 'present situation'. Atlases are vital tools in monitoring the dynamics of organisms within cities. The wise management of natural resources in cities (and indeed elsewhere) for the benefit of people as a whole depends on knowing what organisms are where, how many there are and what changes have occurred.

It is a sad fact that atlases are rarely, if ever, used (let alone effectively) by any of the multiplicity of professions (including biologists) and others who are involved in the planning, design and management of cities. There are many reasons why this is the case—lack of knowledge of them and their value, difficulties in understanding and interpretation, and changes in taxonomy. By way of an example of the latter, many years ago the complier had a 'difficult' discussion with a local authority planning officer and a lawyer about the presence of what they considered to be a rare vascular plant and its importance in determining an application for planning permission. The argument was eventually resolved when both reluctantly accepted that the species was a common taxon that appeared with a different name in an out-of-date flora and atlas that had been superseded. It was not helped by the fact that the distribution of the taxon had changed as the result of boundary changes and greater survey effort. Even so it took a lot of effort to persuade them that the 'two' species were actually the same taxon and that one dot (of a grid map) may represent any number of individuals of the species.

The critical issues that have emerged from the compilation of this bibliography, in no particular order, are as follows:

1. What the bibliography shows, even allowing for its failings, is just how little is known about the organisms present, their distribution and their dynamics in European cities. The point applies even if there are, for example, only 25,000 cities and the number of published atlases is quadrupled.
2. Regrettably the bibliography, which started life 5 or more years ago, is likely to be incomplete and probably inconsistent at least in parts. It is worth reiterating the plea made by Clemants in 2002 (more than a decade ago)—*I welcome any additions or corrections to this list.*
3. Atlases and accounts of the flora and fauna of cities should be related to four distinct geographical units. First, within the limits of the developed area. Second, between the limits of the developed area and the boundary of the geo-political unit. Third, the area between the boundary of the administrative/political unit and the recording area. Fourth, which is a variation of units 2 and 3, the area between the administrative unit and say 20 km from it—this is an arbitrary distance; it could be 15 or 25 km; the purpose is to provide information in relation to the expansion of the urban area and as a base line for measuring changes.
4. To be comparable; the same (or at least similar) methods used for collecting and analysing the data are used for the preparation of an atlas.
5. The national atlases should be merged to produce European atlases.
6. Revisions to atlases should be published every 5 years. In summary the preparation of atlases should be a continuous process.
7. Central governments, city councils and their agencies should allocate sufficient funds for the preparation and collection of atlases; see item 8 below. As to the cost, see items 10 and 11.
8. It is impractical to prepare and publish atlases or the accounts of the flora and fauna of every European city and every 'group' of organisms. Therefore the preparation of atlases should be of one of the two types:

   (a) Structured

   A structured European-wide research programme to prepare and continually update atlases/flora/faunas for several cities of different sizes in each biogeographic region.

   (b) Ad hoc

   The preparation of atlases/floras/fauna on an ad hoc basis throughout Europe.

9. The implementation of an important suggestion made by George Barker of English Nature (now Natural England) at a conference in Poland in 1992, namely that each European country should carry out a critical review of the knowledge about urban areas. This implies the preparation of a bibliography on all aspects (past and present) of the ecology of cities, including atlases. That was more than 20 years ago and it has never been implemented—it needs to be and as a matter of urgency.

10. For many decades there has been a need for a European Centre for the Management of Natural Resources in Cities to initiate, co-ordinate and publish scientific and related information about the planning, design and management of cities in relation to natural resources. The cost of such a centre, the continual preparation of atlases, the production of bibliographies and many other projects are likely to be well within the contingency budgets of European and national military and space research and the funds allocated to particle and theoretical physics, none of which is relevant to the present human predicament. To be somewhat melodramatic about it—urban areas are biological time bombs, a situation that is unappreciated by politicians, their officials and other professions.
11. Animals carry (directly or via other animals) viruses, fungi, bacteria and protozoans that have the potential of devastating human populations and/or other organisms on which they rely but no one knows where they are and how their distribution has changed. The Hubble telescope, the Cern accelerator, the International Space Station and the development of destructive military hardware are not going to be of any help at all, but what is, is what is present, where and how much there is of it.

Chapter 2
Europe

Earth Sciences

Geology

Ager DV (1980) Geology of Europe. McGraw Hill, London
1:5 million scale Geological Map of Europe and Adjacent Areas (c. 2005). Bundesanstadt für Geowissenschafen und Rohstaffe

Geomorphology

Emberton Clifford (ed) (1984) Geomorphology of Europe. Macmillan Press, London

Soils

Kubiena WL (1953) Soils of Europe
Soil Atlas of Europe (2008) European Commission

Major Habitats/Vegetation Types

Forests/Woodlands

No publications found.

Gardens, Parks and Industrial Habitats

No publications found.

Grasslands

No publications found.

Heathland

No publications found.

Marine

No publications found.

© Springer International Publishing Switzerland 2016

J.G. Kelcey, *Provisional Bibliography of Atlases, Floras and Faunas of European Cities: 1600–2014*, DOI 10.1007/978-3-319-31120-3_2

Scrub

No publications found.

Uplands

No publications found.

Water, Moving (Lotic)

Tockner K, Robinson CT, Uehlinger U (eds) (2009) Rivers of Europe. Academic Press, London
Whitton BA (1984) Ecology of European Rivers. Blackwell, Oxford

Water, Still (Lentic)

No publications found.

Wetlands

No publications found.

Animals (Invertebrates)

Acari (Mites and Ticks)

No publications found.

Annelida (Worms)

No publications found.—also see Oligochaeta and Platyhelminthes.

Anoplura (Lice)

No publications found.

Arachnida (Spiders and Harvestmen)

No publications found.

Chilopoda (Centipedes)

See Myriapoda.

Coleoptera (Beetles)

No publications found.

Collembola (Springtails)

Gisin H (1960) Collembolenfauna Europas. Mus Hist Nat Geneve

Decapoda (Crayfish, Shrimps, Crabs and Related Taxa)

Souty-Grasset C, Holdich DM, Noel PY, Reynolds JD, Haffner P (2006) Atlas of the Crayfish in Europe. Pub. Museum national 'Histoire naturalle, Paris

Dermaptera (Earwigs)

Harz K (1969) The Orthoptera of Europe Vol. III W Junk, The Hague

Dictyoptera (Cockroaches)

Harz K (1969) The Orthoptera of Europe Vol. III W Junk, The Hague

Diplopoda (Millipedes)

See Myriapoda.

Diptera (Flies)

No publications found.

Ephemeroptera (Mayflies)

No publications found.

Hemiptera-Heteroptera/Homoptera (Bugs)

No publications found.

Hymenoptera (Ants, Bees, Wasps and Related Taxa)

No publications found.

Isopoda (Woodlice)

No publications found.

Isoptera (Termites)

Harz K (1969) The Orthoptera of Europe Vol. III W Junk, The Hague

Lepidoptera (Butterflies and Moths)

Kudrna O (ed) (1985, 1986) Butterflies of Europe, Vols. 1–8 Aula-Verlag, Weisbaden

Svendson P, Fibiger M (1992) The Distribution of European Macrolepidoptera Faunistica Lepidopterorum Europaerum. European Faunistical Press, Copenhagen

Karsholt O, Razoski J (1996) The Lepidoptera of Europe; a distributional checklist. Apollo Books, Stenstrup

Tolman T, Lewington R (1997) Collins Field Guide, Butterflies of Britain and Europe. HarperCollins, London

Lafranchis T (2004) Butterflies of Europe; a new field guide and key. Diatheo, Paris

Mollusca (Molluscs)

No publications found.

Myriapoda (Centipedes, Millipedes)

Kime RD (1990) A provisional atlas of European myriapods. Part 1.—Fauna Europaea Evertebrata 1: 1–109

Kime RD, Enghoff H (2011) Atlas of the European Millipedes (Class Diplopoda). Vol. 1. Orders Polyxenida, Glomerida, Platydesmida, Siphonocryptida, Polyzoniida, Callipodida, Polydesmida. Pensoft, Sofia-Moscow

Bonato L, Minelli A (2014) Chilopoda Geophilomorpha of Europe: a revised list of species, with taxonomic and nomenclatorial notes. Zootaxa, (S.l.), v. 3770, n. 1, pp 1–136

Mysidacea (Shrimps)

No publications found.

Neuroptera (Lacewings)

Aspock H, Aspock U, Holzel H (1980) Die Neuropteren Europas Vols. I and II (The Neuroptera of Europe). Goecke & Evers, Krefeld

Odonata (Dragonflies and Damselflies)

d'Aguilar J, Dommanget J-L, Prechac R (1986) A Field Guide to the Dragonflies of Britain, Europe and North Africa. Collins, London

Askew RR (1988) The Dragonflies of Europe. Harley Books, Colchester

Oligochaeta (Segmented Worms)

No publications found.

Orthoptera (Grasshoppers and Crickets)

Harz K (1969) The Orthoptera of Europe Vols. I and II. W Junk The Hague

Phasmida (Stick Insects)

Harz K (1969) The Orthoptera of Europe, Vol. III. W Junk, The Hague

Platyhelminthes (Flatworms)

No publications found.

Plecoptera (Stoneflies)

No publications found.

Siphonoptera (Fleas)

No publications found.

Animals (Vertebrates)

Amphibia (Amphibians)

Böhme W (Hrsg.) (c. 1981-2008) Handbuch der Reptilien und Amphibien Europas. (Handbook of the Reptiles and Amphibians of Europe) AULA Wiesbaden. Note: Precise date and volume details not known.

Gasc J-P, Cabela A, Crnobrnja-Isailovic J, Dolmen D, Grossenbacher K, Haffner P, Lescure J, Martens H, Martinez Rica JP, Maurin H, Oliveira ME, Sofianidou TS, Veith M, Zuiderwijk A (eds) (1997) Atlas of Amphibians and Reptiles in Europe. Societas Europaea Herpetologica & Musem National d'Histoire Naturella, Paris

Arnold EN, Burton JA, Ovenden DW (1978). A Field Guide to the Reptiles and Amphibians of Britain and Europe. HarperCollins, London

Mertens R, Muller L (1940) Die Amphibien und Reptilien Europas (The amphibians and reptiles of Europe)—Abh Senckenb Naturf Ges 45(1): 1–56

Mertens R, Wermuth H (1960) Die Amphibien und Reptilien Europas (The amphibians and reptiles of Europe)—Frankfurt am Main

Aves (Birds)

Voous KH (1960) Atlas of European Birds. Nelson, London

Cramp S, Simmons KEL (eds) (1977) Handbook of the Birds in Europe, the Middle East and North Africa. Vols. 1–9 Oxford University Press, Oxford

Peterson R, Mountfort G, Hollom PAD (1993) Collins Field Guide to the Birds of Britain & Europe, 5th edn. HarperCollins, London

Hagemeijer EJM, Blair MJ (eds) (1998) The European Bird Census Council Atlas of European Breeding Birds; their distribution and abundance. T& A.D. Poyser, London

Snow DW, Perrins CM (eds) (1998) Birds of the Western Palearctic (Concise Edition 2 vols.) Oxford University Press, Oxford

Huntley B, Green RE, Collingham YC, Willis SG (2007) A Climatic Atlas of European Breeding Birds. Lynx Editions, Barcelona

Mammalia (Mammals)

Corbet GB (1978) The Mammals of the Palaearctic Region—a Taxonomic Review. British Museum (Natural History), London

MacDonald D, Barrett P (1993) Collins Field Guide, Mammals of Britain and Europe. HarperCollins, London

Mitchell-Jones A, Amori G, Bogdanowicz W, Krystufek B, Reijnders PJH, Spitzenberger F, Stubbe M, Thissen JBM, Vohralik V, Zima J (1999) The Atlas of European Mammals. T & A D Poyser, London

Temple HJ, Terry A (2007) The status and distribution of European Mammals. Luxembourg Office for Official Publications of the European Communities

Aulagnier S, Haffner P, Mitchell Jones AJ, Moutou F (2015) Mammifères d'Europe, d'Afrique du Nord et du Moyen-Orient (The Mammals of Europe, North Africa and the Middle East) (French). Publisher Delachaux et Niestlé. Note: Included because it appears to be an up to date version of Mitchell Jones—1999, see above

Pisces (Fish)

Lelek A (1987) The Freshwater Fishes of Europe (9 vols.). Aula Verlag, Wiesbaden

Miller PJ, Loates MJ (1997) Collins Pocket Guide, Fishes of Britain and Europe. HarperCollins, London

Kottelat M, Freyhof J (2007) Handbook of European Freshwater Fishes. Kottelat, Cornol, Switzerland and Freyhof, Berlin, Germany

Reptilia (Reptiles)

See Amphibians.

Bacteria and Related Organisms

There appears to be no accounts of the European distribution and status of bacteria and related organisms.

Fungi

(a) Excluding Lichenised Fungi

Garnweidner E (2001) Collins Nature Guide, Mushrooms and Toadstools of Britain and Europe. HarperCollins, London

Jordan M (2004) The Encyclopedia of Fungi of Britain and Europe, Frances Lincoln

(b) Lichenised Fungi

No publications found.

Plants (Non-vascular)

Algae

No publications found.

Bryophyta (Bryophytes)

Frey W, Frahm, J-P, Fischer E, Lobin W (2006) The Liverworts, Mosses and Ferns of Europe. Harley Books, Colchester

Plants (Vascular)

Tutin TG, Heywood, VH, Burges NA, Moore DM, Valentine DH, Walters SM, Webb DA (eds) (1964–1980) Flora Europaea Vols. 1–5 Cambridge University Press, Cambridge

Tutin TG, Burges NA, Chater AO, Edmondson JR, Heywood, VH, Moore DM, Valentine DH, Walters SM, Webb DA (eds) (1993) Flora Europaea Vol. I (second edition) Cambridge University Press, Cambridge. Note: It appears that there is only a second edition of Vol. 1 and not Vols. 2–5.

Atlas Florae Europaeae (1972–2014) Distribution of Vascular Plants in Europe Vols. 1–16 (Vols. 17 and 18 in press). Committee for Mapping the Flora of Europe, Helsinki

Protozoans

No publications found.

Viruses

No publications found.

Chapter 3
Animals (Invertebrates)

Acari (Ticks and Mites)

Poland

Warszawa (Warsaw)

Niedbala W, Blaszak C, Blosayk J, Kaliszewski M, Kazmierski A (1982) Soil mites (Acari) of Warsaw and Mazovia. In: Species Composition and Origin of the Fauna of Warsaw Part 3, Memorabilia Zoologica 36: 235–251

Anoplura (Lice)

England

London

Girling MA (1984) Eighteenth Century Records of Human Lice (Pthiraptera, Anoplura) and Fleas (Siphonaptera, Pulicidae) in the City of London. Ent. Monthly Magazine 120: 207–210

Arachnids (Spiders and Harvestmen)

© Springer International Publishing Switzerland 2016
J.G. Kelcey, *Provisional Bibliography of Atlases, Floras and Faunas of European Cities: 1600–2014*, DOI 10.1007/978-3-319-31120-3_3

Photo: John G. Kelcey

Czech Republic

Praha (*Prague*)

Kurka A, Buchar J, Rezac M (2007) Pavouci (Araneae) Prahy (Spiders, Araneae) of Prague (Czech Republic). Natura Pragensis 18: 5–126

England

London

Savory TH, Le Gros AE (1957) The Arachnida of London. London Naturalist 36: 41–50

Milner JE (1995) Spider Records for the London Area in 1994. London Naturalist 74: 139–143

Milner JE (1997) Spider Records for the London Area in 1996. London Naturalist 76: 153–156

Milner JE (1998) Spider Records for the London Area in 1997. London Naturalist 77: 147–148

Milner JE (2000) Spiders in London. London Naturalist 79: 117–142

Sheffield

Roberts DJ, Roberts MJ (1985) Spiders. In: Whiteley D (ed) The Natural History of the Sheffield Area. Sorby Natural History Society: 210–221

Poland

Wasrzawa (*Warsaw*)

Krzyanowska, E, Dziabaszewski A, Jackowska B, Starega W (1981) Spiders (Arachnoidea, Arnae). In: Species Composition and Origin of the Fauna of Warsaw Part 1, Memorabilia Zoologica 34: 87–110

Czechowska W, Kubicka A, Starega W (1981) Harvestmen (Arachnoidea, Opiliones) of Warsaw and Mazovia. In: Species Composition and Origin of the Fauna of Warsaw Part 1, Memorabilia Zoologica 34: 111–118

Chilopoda (Centipedes)

See Myriapoda.

Coleoptera (Beetles)

Photo: Lubomir Hlašek

Bulgaria

Sofia

Stoyanov IL (2004) The Ground Beetle (Coleoptera: Carabidae) Fauna of Sofia, Bulgaria. A checklist. In: Penev Stoyanov IL, Niemela J, Kotze, DJ and Chipev, N (eds) (2004) Ecology of the City of Sofia. Pensoft Sofia-Moscow, pp 401–415

Czech Republic

Praha (*Prague*)

Vesevly P (2002) Strevlikoviti brouci Prahy (Coleoptera: Caraabidae) (Ground Beetles of Prague). Clairon Production, Praha (In Czech and German with English abstract)

England

London

Mabbott P (2000) Coleoptera: Carabidae in the London Area with a preliminary list of the ground beetles. London Naturalist 79: 165–176

Sheffield

Lazenby A (1988) Urban Beetles in Sheffield. Sorby Record 25: 22–31

Poland

Warszawa (*Warsaw*)

Burakowski B, Nowakowski E (1981) Click Beetles (Coleoptera, Elateridae) of Warsaw and Mazovia. In Species Composition and Origin of the Fauna of Warsaw Part 1, Memorabilia Zoologica 34: 165–180

Burakowski B, Nowakowski (1981) Longicorns (Coleoptera, Cerambycidae) of Warsaw and Mazovia. In Species Composition and Origin of the Fauna of Warsaw Part 1, Memorabilia Zoologica 34: 199–218

Cholewicka K (1981) Curculionids (Coleoptera, Curculionidae) of Warsaw and Mazovia. In Species Composition and Origin of the Fauna of Warsaw Part 1, Memorabilia Zoologica 34: 235–260

Czechowska W, Bielawaski R (1981) Coccinellids (Coleoptera, Coccinellidae) of Warsaw and Mazovia. In Species Composition and Origin of the Fauna of Warsaw Part 1, Memorabilia Zoologica 34: 187–197

Czechowska W (1981) Carabids (Coleoptera, Carabidae) of Warsaw and Mazovia In Species Composition and Origin of the Fauna of Warsaw Part 1, Memorabilia Zoologica 34: 119–144

Kubicka A (1981) Scarabaeids (Coleoptera, Scarabaeidae) of Warsaw and Mazovia. In Species Composition and Origin of the Fauna of Warsaw Part 1, Memorabilia Zoologica 34: 145–164

Wasowska M (1981) Chrysomelids (Coleoptera, Chrysomelidae) of Warsaw and Mazovia. In: Species Composition and Origin of the Fauna of Warsaw Part 1, Memorabilia Zoologica 34: 219–233

Slovakia

Banska Bystrica

Franc V (1992) (Title in Slovak?) Interesting findings of beetles (Coleoptera) in the urban area of Banska Bystrica (Slovakia). Acta Universitatis Carolinae Biologica 35: 229–237

Collembola (Springtails)

Poland

Warszawa (*Warsaw*)

Sterzynska M (1982) Springtails (Collembola) of Warsaw and Mazovia. In Species Composition and Origin of the Fauna of Warsaw Part 3, Memorabilia Zoologica 36: 217–234

Diplopoda (Millipedes)

See Myriapoda.

Diptera (Flies)

Photo: M.A.J. Grutters

England

Bristol

Audcent HLF (1948) Bristol Insect Fauna (Diptera). Proc. Naturalists' Society XXVII, part V 1948: 406–470. Published 1949

Audcent HLF (1949) Bristol Insect Fauna (Diptera). Proc. Naturalists' Society XXVIII, part V 1948: 45–132. Published 1950

London

Parmenter L (1952) The Flies of the London Area I Conopidae. London Naturalist 31: 25–29

Nye ER (1955) The Flies of the London Area II Culicidae, Sub-family Culicinae (Mosquitoes) with a key to the species of *Culex*. London Naturalist 34: 114–126

Niblett M (1956) Flies of the London Area III, Trypetidae. London Naturalist 35: 82–88

Plant CW (1986) A Working List of the Hoverflies (Diptera: Syrphidae) of the London Area. London Naturalist 65: 109–117

Plant CW (1990) Hoverflies (Diptera: Syrphidae) of the London Area: Progress Report and Selected Distribution Maps. London Naturalist 69: 53–66

Sheffield

Whiteley D (1987) Hoverflies of the Sheffield Area. Sorby Record Special Series 6: 1–56

Poland

Warszawa (Warsaw)

Bankowska R (1982) Dolichopodidae (Diptera) of Warsaw and Mazovia. In: Species Composition and Origin of the Fauna of Warsaw Part 2, Memorabilia Zoologica 35: 3–31

Bankowska R (1982) Hoverflies (Diptera, Syrphidae) of Warsaw and Mazovia. In: Species Composition and Origin of the Fauna of Warsaw Part 2, Memorabilia Zoologica 35: 57–78

Bankowska R (1982) Conopidae (Diptera) of Warsaw and Mazovia. In: Species Composition and Origin of the Fauna of Warsaw Part 2, Memorabilia Zoologica 35: 79–83

Bankowska R (1982) Pipunculidae (Diptera) of Warsaw and Mazovia. In: Species Composition and Origin of the Fauna of Warsaw Part 2, Memorabilia Zoologica 35: 85–91

Draber-Monko A (1982) Scatophagidae (Diptera) of Warsaw and Mazovia. In: Species Composition and Origin of the Fauna of Warsaw Part 2, Memorabilia Zoologica 35: 115–121

Draber-Monko A (1982) Calliphoridae parasitica (Diptera) of Warsaw and Mazovia. In: Species Composition and Origin of the Fauna of Warsaw Part 2, Memorabilia Zoologica 35: 123–129

Draber-Monko A (1982) Sarcophagidae and Rhinophoridae (Diptera) of Warsaw and Mazovia. In: Species Composition and Origin of the Fauna of Warsaw Part 2, Memorabilia Zoologica 35: 131–140

Draber-Monko A (1982) Tachinid flies (Diptera, Tachinidae) of Warsaw and Mazovia. In: Species Composition and Origin of the Fauna of Warsaw Part 2, Memorabilia Zoologica 35: 141–162

Draber-Monko A (1982) Gasterophilida, Hypodermatidae, Oestridae, Hippoboscidae and Nycteribiidae (Diptera) of Warsaw and Mazovia. In: Species Composition and Origin of the Fauna of Warsaw Part 2, Memorabilia Zoologica 35: 163–168

Durska E (1982) Phoridae (Diptera) of Warsaw. In: Species Composition and Origin of the Fauna of Warsaw Part 2, Memorabilia Zoologica 35: 47–56

Gorska D (1982) Anthomyiidae, Muscidea and non-parasitic Calliphoridae (Calyptrata, Diptera) of Warsaw and Mazovia. In: Species Composition and Origin of the Fauna of Warsaw Part 2, Memorabilia Zoologica 35: 93–114

Trojan P (1982) Diptera (Tabanomorpha) of Warsaw and Mazovia. In: Species Composition and Origin of the Fauna of Warsaw Part 2, Memorabilia Zoologica 35: 3–31

Wegner E (1982) Mosquitoes (Diptera, Culicidae) of Warsaw and Mazovia. In: Species Composition and Origin of the Fauna of Warsaw Part 3, Memorabilia Zoologica 36: 201–216

Hemiptera (Bugs)

Photo: Lubomir Hlašek

Bulgaria

Sofia

Tasheva-Terzieva E (2004) A checklist of the Aphids (Homoptera, Aphidodea) from the City of Sofia. In: Penev Stoyanov IL, Niemela J, Kotze DJ, Chipev N (eds) (2004) Ecology of the city of Sofia. Pensoft Sofia-Moscow, pp 365–370

England

Hitchin

Palmer R (1934) Hemiptera. In: Hine RL (ed) The Natural History of the Hitchin Region. Hitchin and District Regional Survey Association, Hitchin, pp 113–116

London

Groves EW (1964) Hemiptera-Heteroptera of the London Area. Part I. London Naturalist 43: 34–66

Groves EW (1965) Hemiptera-Heteroptera of the London Area. Part II. London Naturalist 44: 82–110

Groves EW (1966) Hemiptera-Heteroptera of the London Area. Part III. London Naturalist 45: 60–88

Groves EW (1967) Hemiptera-Heteroptera of the London Area. Part IV. London Naturalist 46: 82–104

Groves EW (1968) Hemiptera-Heteroptera of the London Area. Part V. London Naturalist 47: 50–80

Groves EW (1969) Hemiptera-Heteroptera of the London Area. Part VI. London Naturalist 48: 86–120

Groves EW (1972) Hemiptera-Heteroptera of the London Area. Part VII. London Naturalist 50: 87–94

Groves EW (1973) Hemiptera-Heteroptera of the London Area. Part VIII. London Naturalist 52: 31–59

Groves EW (1975) Hemiptera-Heteroptera of the London Area. Part IX. London Naturalist 54: 21-34

Groves EW (1976) Hemiptera-Heteroptera of the London Area. Part X. London Naturalist 55: 6–15

Groves EW (1977) Hemiptera-Heteroptera of the London Area. Part XI. London Naturalist 56: 32–43

Groves EW (1982) Hemiptera-Heteroptera of the London Area. Part XII. London Naturalist 61: 72–87

Groves EW (1983) Hemiptera-Heteroptera of the London Area. Part XIII. London Naturalist 62: 69–86

Groves EW (1984) Hemiptera-Heteroptera of the London Area. Part XIV. London Naturalist 63: 97–120

Groves EW (1985) Hemiptera-Heteroptera of the London Area. Part XV. London Naturalist 64: 63–94

Groves EW (1986) Hemiptera-Heteroptera of the London Area. Part XVI. London Naturalist 65: 119–180

Sheffield

Zasada KA, Smith EH (1981) Freshwater Invertebrates of the Sheffield District. Sorby Record Special Series 4

Note: This publication appears here on the assumption that it is mainly concerned with aquatic Heteroptera although the title indicates that it includes other Orders and Phyla such as the Ephemoptera, Plecoptera, Trichoptera and Mollusca.

Poland

Warszawa (*Warsaw*)

Chudizicka E (1982) Auchenorrhymcha (Homoptera) of Warsaw and Mazovia. In: Species Composition and Origin of the Fauna of Warsaw Part 3, Memorabilia Zoologica 36: 143–164

Heteroptera (including Shield, Ground and Water Bugs)

See Hemiptera.

Homoptera (Cicadas, Froghoppers, Leafhoppers and Aphids)

See Hemiptera.

Hymenoptera (Ants, Bees and Wasps)

Photo: M.A.J. Grutters

<u>Bulgaria</u>

Sofia

Lapeva-Gjonova A, Atanasova M (2004) A Faunistic Investigation of the Ants (Hymenoptera, Formicidae) in the city of Sofia. In: Penev Stoyanov IL, Niemela J, Kotze, DJ, Chipev, N (eds) (2004) Ecology of the City of Sofia. Pensoft Sofia-Moscow, pp 401–422

<u>Finland</u>

Helsinki

Backman JPC, Teras I (1999) Bumblebees (Hymenoptera, Apidae) in the city of Helsinki. Proceedings of the XXIV Nordic Congress of Entomology, pp 11–22

<u>Poland</u>

Warszawa (*Warsaw*)

Banaszak J (1982) Apoidea (Hymenoptera) of Warsaw and Mazovia. In: Species Composition and Origin of the Fauna of Warsaw Part 3, Memorabilia Zoologica 36: 129–142

Garbarczyk H (1982) Proctotrupoidea (Hymenoptera) of Warsaw and Mazovia. In: Species Composition and Origin of the Fauna of Warsaw Part 3, Memorabilia Zoologica 36: 41–60

Pisarski B (1982) Ants (Hymenoptera, Formicoidea) of Warsaw and Mazovia. In: Species Composition and Origin of the Fauna of Warsaw Part 3, Memorabilia Zoologica 36: 73–90

Sawoniewicz J (1982) Ichneumonidae (Hymenoptera) of Warsaw and Mazovia. In: Species Composition and Origin of the Fauna of Warsaw Part 3, Memorabilia Zoologica 36: 5–39

Skibinska E (1982) Chrysididae (Hymenoptera) of Warsaw and Mazovia. In: Species Composition and Origin of the Fauna of Warsaw Part 3, Memorabilia Zoologica 36: 201–216

Skibinska E (1982) Wasps (Hymenoptera, Vespidae) of Warsaw and Mazovia. In: Species Composition and Origin of the Fauna of Warsaw Part 3, Memorabilia Zoologica 36: 91–102

Skibinska E (1982) Sphecidae (Hymenoptera) of Warsaw and Mazovia. In: Species Composition and Origin of the Fauna of Warsaw Part 3, Memorabilia Zoologica 36: 103–127

Isopoda (Woodlice)

Photo: Lubomir Hlašek

Poland

Warszawa (*Warsaw*)

Jedryczkowski W (1981) Isopods (Isopoda) of Warsaw and Mazovia. In: Species Composition and Origin of the Fauna of Warsaw Part 1, Memorabilia Zoologica 34: 79–86

Lepidoptera (Butterflies and Moths)

Photo: John G. Kelcey

Bulgaria

Razgrad

Markowitsch A (1900) Beitrag zu den Lepidoptera der Umgebung von Razgrad. (Contribution to the Lepidoptera around Razgrad). Svetulka 1(5): 34–35

Sliven

Tschorbadjiew P (1919) Beitrag zur Schmetterlingsfauna der Stadt Silven und ihrer Umgebung. (Contribution to the butterfly fauna of Silven and its environs). Z Bulg Akad Wiss 17: 175–214

Swilengrad (Svilengrad)

Tschorbadjiew P (1928) Beitrag zur Lepidopterenfauna der Umgebung von dem Bezirk Swilengrad. (Contribution to Lepidopteran fauna around the area of Svilengrad). Arb Bulg Nat Ges 13: 173–182

Tirnovo (Turnovo)

Tschorbadjiew P (1930) Erster Beitrag zur Lepidopterenfauna der Stadt Tirnovo und ihre Umgebung. (First contribution to the Lepidopteran fauna of Turnovo and its vicinity). Mitt Bulg Ent Ges 5: 125–162

England

Aldermaston

Eastwick-Field GG (1983) Macro-lepidoptera of Aldermaston 1976-1982. ARE Conservation Group

Berkhamstead

Hodgson SB (1936) Notes on the Lepidoptera of the Berkhamsted District. Transactions Hertfordshire Natural History Society 20: 41–43

Bishops Stortford

Anon (1950) List of the Butterflies of Bishop's Stortford. Bishop's Stortford Natural History Society 1: 3–9

Anon (1985). List of the Macro-lepidoptera of Bishop's Stortford. Bishop's Stortford Natural History Society

Bracknell

Dumbleton M (2003) The Moths of Bracknell, Bracknell 1976-1985. Reading and District Natural History Society

Brentwood

Williams EF (1953) Butterflies in the Brentwood Area in 1952. Essex Naturalist 29: 110–111

Bristol

Barnett R, Higgins R, Moulin T, Wiltshire C (2003) Butterflies of the Bristol Region. Bristol Regional Environmental Records Centre, Bristol

Barnett RJ, Andrews RM, Bailey MA, Corner T, Martin JP (2008) Moths of the Bristol Region. Bristol Regional Environmental Records Centre, Bristol

Chalfont

Ansorge E (1961) Macro-lepidoptera of the Chalfont area of Buckinghamshire. Entomologist's Gazette qw: 11–18

Cheshunt

Boyd WC (1901) List of the Lepidoptera of Cheshunt and its neighbourhood. Transactions Hertfordshire Natural History Society 11: 75–86

Doncaster

Rimington E (1992) Butterflies of the Doncaster District. Sorby Record Special Series 9. Sorby Natural History Society

Formby

Leech MJ, Michaelis HN (1957) The Lepidoptera of Formby. Raven Entomological and Natural History Society

Greater Manchester

Hardy PB (1998) Butterflies of Greater Manchester. PGL Enterprises, Sale

Halifax

Collinson WE (1969) The Butterflies and Moths of Halifax and District. Halifax Scientific Group

Harrogate

Barnham M, Foggitt GT (1987) Butterflies in the Harrogate District. Harrogate and District Naturalists' Society. Publisher also given as M Barnham and CT Foggitt

Hatfield

Hodgson LS (1939) Notes on the Lepidoptera of the Hatfield District in 1937 and 1938. Transactions Hertfordshire Natural History Society 21: 69–70

Hitchin

Durrant JH (1885) List of Lepidoptera observed in the neighbourhood of Hitchin and at Knebworth. Transactions Hertfordshire Natural History Society 3: 261–266

Foster AH (1934) Butterflies and Moths. In: Hine RL (ed) The Natural History of the Hitchin Region. The Hitchin and District Regional Survey Association, Hitchin

Letchworth

Bowden SR (1963) Butterflies and other insects. In: Ross DA (ed) In and around Letchworth. Letchworth Naturalists' Society

Leyton (now in Greater London)

Meldola R (1891) The Lepidoptera of Leyton and neighbourhood, a contribution to the county fauna. Essex Naturalists 5: 153–170

Liverpool

Byerley I (1854) The Fauna of Liverpool. Appendix to the Proceedings of Liverpool Literary and Philosophical Society, 1853–1854

Anon (1987) Butterflies of Merseyside. Merseyside Branch of the British Butterfly Conservation Society

London

Buckell MB, Prout LB (1899-1902) The Fauna of the London District: Lepidoptera. Transactions, London Ent. Nat. Hist Soc. 8: 51–63; 9: 66–80; 10: 62–74; 11: 54–68.

de Worms CGM (1950) The Butterflies of London and its surroundings. London Naturalist 29: 46–80

de Worms CGM (1954) The Moths of London and its surroundings. London Naturalist 33: 101–146

de Worms CGM (1955) The Moths of London and its surroundings. London Naturalist 34: 66–107

de Worms CGM (1956) The Moths of London and its surroundings. London Naturalist 35: 33–76 de Worms CGM (1957) The Moths of London and its surroundings. London Naturalist 36: 59–99

de Worms CGM (1958) The Moths of London and its surroundings. London Naturalist 37: 136–178

de Worms CGM (1959) A Supplement to the Butterflies of London and its vicinity. London Naturalist 38: 33–50

de Worms CGM (1960) A supplement to the Butterflies and Moths of London and its vicinity. London Naturalist 39: 99–129

de Worms CGM (1962) A review of the Lepidoptera of the London area for 1960 and 1961. London Naturalist 41: 60–65

de Worms CGM (1964) A review of the Macro-lepidoptera of the London area for 1962 and 1963. London Naturalist 43: 26–33

de Worms CGM (1966) A review of the Macro-lepidoptera of the London area for 1964 and 1965. London Naturalist 45: 89–97

St. John WG (1968) Butterflies in the London Area. Entomologists' Gazette 19: 32

de Worms CGM (1969) Resumé of the Macro-lepidoptera of the London Area for 1966 and 1967. London Naturalist 48: 121–124

de Worms CGM (1970) A review of the Macro-lepidoptera of the London area for 1968 and 1969. London Naturalist 49: 82–89

de Worms CGM (1972) A review of the Macro-lepidoptera of the London area for 1970 and 1971. London Naturalist 51: 28–38

de Worms CGM (1974) A review of the Macro-lepidoptera of the London area for 1972 and 1973. London Naturalist 53: 86–99

de Worms, CGM (1976) A review of the Macro-lepidoptera of the London Area for 1974 and 1975. London Naturalist 55: 48–57.

de Worms, CGM (1978) A review of the Macro-lepidoptera of the London Area for 1976 and 1977. London Naturalist 57: 81–90

Plant CW (1981) A review of the Lepidoptera of the London Area 1978-1980. London Naturalist 60: 100-104

Plant CW (1983) A review of the Lepidoptera of the London Area for 1981 and 1982. London Naturalist 62: 116–125

Plant CW (1987) A review of the Lepidoptera of the London Area for 1985 and 1986. London Naturalist 66: 135–156

Plant CW (1989) A review of the Lepidoptera of the London Area for 1987 and 1988. London Naturalist 68: 89–107

Plant CW (1993) The Larger Moths of the London Area. London Natural History Society

Plant CW (1997) A review of the Butterflies and Moths (Lepidoptera) of the London Area for 1995-1996. London Naturalist 76: 157–174

Plant CW (1999) A review of the Butterflies and Moths (Lepidoptera) of the London Area for 1997-1998. London Naturalist 78: 147–171

Newbury

Saundy RMS (1967) The Macro-lepidoptera of the Newbury District. Transactions Newbury and District Field Club 11: 31–59

Oxford

Waters EGR (1929) A List of the Micro-lepidoptera of the Oxford District. Proceedings Ashmolean Natural History Society: 1928

Bretheron RF (1940) A list of the Macro-lepidoptera of the Oxford District. Proceedings Ashmolean Natural History Society: 1939

Bretheron RF (1941) Additions to the list of the Macro-lepidoptera of the Oxford District. Proceedings Ashmolean Natural History Society: 1940

Emmet AM (1948) Second supplement to the Macro-lepidoptera of the Oxford District. Proceedings Ashmolean Natural History Society

Note: Oxford is likely to have spread into the adjacent 'District' since 1929.

Scarborough

Walsh GB (1956) Order Lepidoptera. In: Walsh GB, Rimington FC (eds) Natural History of the Scarborough District II, Scarborough. Scarborough Field Naturalist's Society

Sheffield

Fearnehough TD (1938) Butterflies of the Sheffield District. Entomologist's Record. J Var 50: 83–84

Garland SP (1979) The Moths of Sheffield. Sorby Record. Special Series 1. Sorby Natural History Society

Garland SP (1981) Butterflies of the Sheffield Area. Sorby Record Special Series 5, Sorby Natural History Society

Garland S Butterflies and Moths (Lepidoptera) (1985). In: Whiteley D (ed) (1985) The Natural History of the Sheffield Area. Sorby Natural History Society, Sheffield, pp 135–151

Whiteley D (ed) (1992) Butterfly Atlas. Sorby Record: 29, Sorby Natural History Society, Sheffield

Watford

Penrose RJ (1980) Butterflies and Moths at Watford 1948-1962. Transactions Hertfordshire Natural History Society 28: 92–100

Poland

Warszawa (Warsaw)

Winiarska G (1982) Noctuids (Noctuidae, Lepidoptera) of Warsaw and Mazovia. In: Species composition and origin of the fauna of Warsaw. Part 3. Memorabilia Zoologica 36: 185–200

Romania

Bucuresti (Bucharest)

Szekely L (2011) The Macrolepidoptera of Bucharest and its vicinity (Romania). Travaux du Museum d'histoire Naturelle 'Grigore Antipa,' 54(2): 461–512

Mecoptera (Scorpion Flies)

See Neuroptera.

Mollusca (Molluscs)

Photo: Lucie Jurickova

Czech Republic

Hradec Králové

Jurickova L (1998) Měkkýši Hradec Králové (Východní Čechy, Česká republika) (Molluscs of Hradec Králové (East Bohemia, Czech Republic)). Acta mus. Reginaebradecensis 26: 101–172

Praha (Prague)

Jurickova L (1995) Mekkysi fauna Velke Prahy a jeji vyvoj pod vlivem urbanizace (Molluscan fauna in the territory of Greater Prague). Natura Pragensis 12: 1–212

England

Scarborough

Walsh GB, Rimington FC (eds) (1956) The Natural History of Scarborough, Vol. 2 (Zoology)

York

Brooke TA (1906) Mollusca. In: Auden GA (ed) Historical and scientific survey of York and District. British Association for the Advancement of Science, York and London

<u>Russian Federation</u>

Moscow

Schileyko AA (1981) Gastropoda наземного из Московской области, почвы фауна Московской области (Terrestrial gastropods of the Moscow Area, Soil fauna of the Moscow Area) 1: 144–169

Myriapoda (Centipedes and Millipedes)

Photo: Mihaela Ion

<u>Bulgaria</u>

Sofia

Stoev P (2004) Myriopoda (Chilopoda, Diplopoda) in urban environments in the city of Sofia. In: Penev Stoyanov IL, Niemela J, Kotze, DJ and Chipev, N (eds) (2004) Ecology of the City of Sofia. Pensoft Sofia-Moscow

<u>Poland</u>

Warszawa (*Warsaw*)

Jedryczkowski W (1982) Millipedes (Diplopoda) of Warsaw and Mazovia. In: Species Composition and Origin of the Fauna of Warsaw Part 3, Memorabilia Zoologica 36: 253–261

Neuroptera (Lacewings and Scorpion Flies)

Photo: Lubomir Hlašek

England

London

Plant CW (1992) A Working List of the Lacewings of the London Area. London Naturalist 71: 117–136

Poland

Wasrzawa (Warsaw)

Czechowska W (1982) Neuroptera and Mecoptera of Warsaw and Mazovia. In: Species composition and origin of the fauna of Warsaw. Part 3. Memorabilia Zoologica 36: 165–184

Odonata (Dragonflies and Damselflies)

Photo: John G. Kelcey

England

Bristol

Randolph S (1992) Dragonflies of the Bristol Region. City of Bristol Museums and Art Gallery

Chelmsford

Shanahan J (1982) Odonata. In: Nature around Chelmsford. Chelmsford Nat Hist Group, pp 31–34

London

Longfield C (1949) The Dragonflies of the London Area. London Naturalist 28: 90–98

Brooks Stephen J (1989) The Dragonflies (Odonata) of London: the current status. London Naturalist 68: 109–132

Germany

Aschaffenburg

Lenk P (1992) Die Libellen der Stadt Aschaffenburg (The dragonflies of Aschaffenburg). Nachrichten d. Naturwissenschaftlichen Museums d. Stadt Aschaffenburg H. 99 S. 1–24

Berlin

Jahn P (1982) Liste der Libellnarten (Odonata) von Berlin (West) mit Kennzeichnung der ausgestorbenen und gefährdeten Arten (Rote Liste). (List of Dragonfly species (Odonata) of the city of Berlin (West) including data of extinct and threatened species (Red Data List). Landschaftsentwicklung u. Umweltforschung 11: 297–310

Jahn P (198) Liste der Libellen des Landes Berlin (Dragonflies of Berlin). Bestandsentwicklung – Gefährdung – Schutz. Beitrag zum Artenschutzprogramm. Sen f Stadtentw u. Umweltschutz Abt III Berlin

Serbia

Dubrovnik

Adamovic ZR (1967) Odonata collected in Dubrovnik District, Yugoslavia. Dt ent Z 14: 285–302

Oligochaetae (Earthworms)

Photo: John G. Kelcey

Poland

Warszawa (*Warsaw*)

Plisko JD (1959) Lumbricidae Warszawy I okolic (Lumbricidae of Warsaw and surrounding areas) Fragm. Fauna (Warsaw) 8: 247–271

Kasprzak K (1981) Enchytraeids (Oligochaeta, Enchytraeidae) of Warsaw and Mazovia. In: Species Composition and Origin of the Fauna of Warsaw Part 1, Memorabilia Zoologica 34: 59–67

Pilipuik I (1981) Earthworms (Oligochaeta, Lumbricidae) of Warsaw and Mazovia. In: Species Composition and Origin of the Fauna of Warsaw Part 1, Memorabilia Zoologica 34: 69–77

Orthoptera (Grasshoppers and Crickets)

Photo: Pavel Marhoul

England

London

Payne RM (1958) The Distribution of Grasshoppers and Allied Insects in the London Area. London Naturalist 37: 102–115

Oxford

Lucas WJ (1926) Orthoptera. In: Walk JJ (ed) The Natural History of Oxford.

Sheffield

Whiteley D (1981) Grasshoppers, Crickets and Cockroaches (Orthoptera) of the Sheffield Area. Sorby Record 19: 68–75

Germany

Hessen

Ingrisch S (1979) Die Orthopteren und Blattopteren (Insecta: Orthoptera, Dermaptera, Blattoptera) von Hessen (The Orthopteren and Blattopteren (Insecta: Orthoptera, Dermaptera, Blattoptera) of Hessen). Fundortkataster der Bundesrepublik Deutschland (In: Müller P (ed) Teil 13; pp. 99 Saarbrücken & Heidelberg)

Giessen

Ingrisch S (1980) Zur Orthopterenfauna der Stadt Giessen (Hessen) (Saltatoria, Dermaptera and Blattoptera) (Orthopteran fauna of the city of Giessen). Entomologische Zeitschrift 90 Nr. 24 S: 273–280

Pthiraptera (Lice)

See next entry.

Siphonaptera (Fleas)

England

London

Girling MA (1984) Eighteenth century records of human lice (Pthiraptera, Anoplura) and fleas (Siphonaptera, Pulicidae) in the City of London. Entomologists. Monthly Magazine 120: 207–210

Syrphidae (Hoverflies)

See Diptera.

Chapter 4
Animals (Vertebrates)

Amphibia (Amphibians)

Photo: Jiři Vojar

<u>Bulgaria</u>

Plovdiv

Mollov I (2011) Habitat Distribution of the Amphibians and Reptiles in the city of Plovdiv. Bulgaria Biharean Biol 5(1): 25–31

<u>Czech Republic</u>

Brno

Kral B, Pellantova J, Kokes J (1983) Amphibians and reptiles of the Brno urban agglomeration. Folia Zool 32 Nr 1 S.: 51–66

© Springer International Publishing Switzerland 2016

J.G. Kelcey, *Provisional Bibliography of Atlases, Floras and Faunas of European Cities: 1600–2014*, DOI 10.1007/978-3-319-31120-3_4

Prague

Kerous K (1996) Stdue vyskytutrid Amphibia a Reptilia v letech 1986-1993. (A study of the Distribution of the Classes, Amphibia and Reptilia 1986-1993) Natura Pragensis, 13. Prague. Studie o prirode Prahy 13, Praha, Cesky Republika

England

Halifax

Morley I (1965) Vertebrate Fauna of the Halifax Parish. Halifax Scientific Society

London

Fitter RSR (1949) A Check-list of the Mammals, Reptiles and Amphibian of the London area 1900-1949. London Naturalist 28: 98–115

Yalden DW (1965) Distribution of Reptiles and Amphibians in the London Area. London Naturalist 44: 57–69

Yalden DW (1967) Reptiles and Amphibians in the London Area 1965-1966. London Naturalist 46: 68–75

King PW (1979) Amphibia and Reptilia in the London Area 1978. London Naturalist 58: 69–70

King PW (1981) Amphibia and Reptilia in the London Area 1979 and 1980. London Naturalist 60: 96–99

King PW (1983) Amphibia and Reptilia in the London Area 1981 and 1982. London Naturalist 62: 112–115

Langton T (1991) Distribution and Status of Reptiles and Amphibians in the London area. London Naturalist 70: 97–123

Sheffield

Whiteley D (1977) Amphibian Fauna of Sheffield. Sorby Record 15: 36–48

Whiteley D (1988) Amphibians and Reptiles of the Sheffield Area. Sorby Record Special Series 9, Sorby Natural History Society, Sheffield

France

Marseille

Berner L (1955) Amphibiens et reptiles des environs de Marseille. (Amphibians and Reptiles of the Marseille District). Bull. Soc. Linn. Provence 20: 45–46

Germany

Altötting

Zahn A, Steiger S, Petri B, Frsterra G (1991) Amphibienkartierung im Landkreis Altötting (Amphibian mapping in Altötting) – Schriftenr. Bayer. Landesamt f. Umweltschutz 113: 113–117

Baden-Württemberg

Höltzinger J, Schmid G (Hrsg.) (1987) Die Amphibien und Reptilien Baden-Württembergs (The amphibians and reptiles of Baden-Württemberg). – Beih. Veröff. Naturschutz Landachaftspfl. Bad-Württ 41

Bayreuth

Schlumprecht H (1986) Amphibienkartierung im Stadtgebiet Bayreuth (Amphibian mapping in the urban area of Bayreuth) – Schriftenr. Bayer Landesamt f. Umweltschutz 73: 203–205

Berlin

Nessing R (1990) Verbreitungatlas der Amphibien und Reptilien in Berlin, Hauptstadt der DDR. Teil I: Amphibien. Teil II: Reptilien. (Distribution maps of amphibians and reptiles in Berlin, capital of the GDR. Part I: Amphibians. Part II: Reptiles) Berlin

Blankenburg

Geitel (1880/1) Reptilien der Umgebung von Blankenburg/Harz. (Reptiles around Blankenburg /Harz) – Jahresber Ver Naturwiss Braunschweig, Seite 71

Calw

Zimmermann P (o.J.) Amphibien und Reptilien im Landkreis Calw – Der landkreis Calw – Ein Jahrbuch (Amphibians and Reptiles in Calw - The county of Calw - A Yearbook) 8: 115–141

Cottbus

Heym WD, Paepke HJ (1978) Amphibien und Reptilien im Bezirk Cottbus (Amphibians and reptiles in Cottbus). Naturu Landscaft i. Bez Cottbus 1: 41–51

Dusseldorf

Elend A, Glaw F, Schultz P (1986) Die Amphibien und Reptilein der Stadt Düsseldorf (Ein Beitrag zum Biotop-und Artenschutz, Stand 1986) (The amphibians and reptiles of Dusseldorf (A contribution to habitat and species protection, Stand 1986)) – Hrsg: BUND-Kreisgruppe Düsseldorf, Rheinisch-Bergischer Naturschutzverein – Stadverband Düsseldorf

Glaw F, Schultz P (1988) Die Amphibien und Reptilein der Stadt Düsseldorf (The amphibians and reptiles of Dusseldorf) – Jb Feldherpetol 2: 23–45

Elberfeld

Behrens WJ (1884) Die Amphibien und Reptilien der Umgegend von Elberfeld (The amphibians and reptiles in the vicinity of Elberfeld) – J ber Naturwiss V Elberfeld 6: 78–79

Fürth

Heimbucher D (1992) Amphibienkartierung im Landkreis Fürth (Amphibian mapping in Fürth) – Schriftenr. Bayer Landesamt f. Umweltschutz 112: 135–139

Gelsenkirchen

Hamann M, Uthoff HD (1994) Amphibien und Reptilien in Gelsenkichen (Amphibians and Reptiles in Gelsenkirchen) – ZeFldherpetol 1: 103–134

Hamburg

Holst J (1928) Űber die Kriechtiere und Lurche der Umgebung Hamburgs. (The reptiles and amphibians around Hamburg) – Bl Aquar Terrar kd 39: 265

Hof

Schlumprecht H (1992) Amphibien der Stadt Hof – Ergebnisse der Stadtbiotopkartierung (Amphibians of the city of Hof - Results of the mapping) – Schriftenr. Byer. Landesamt f. Umweltschutz 112: 185–190

Magdeburg

Badewitz HJ, Bischoff W, Jaeger D (1967) Die Amphibien und Reptilien Magdeburgs und seiner näheren Umgebung. (The amphibians and reptiles of Magdeburg and its immediate vicinity) – Natursch Naturkdl Heimatfirsch Bez Halle u. Magdeburg 4: 68–72

Mühldorf

Zahn A (1991) Stand der Amphibienkartierung im Landkreis Mühldorf (Status of amphibians mapping in the Mühldorf area) – Schriftenr Bayer Landesamt f. Umweltschutz 113: 107–112

Regensburg

Geissner W (1992) Amphibienkartierung im Landkreis Regensburg (Amphibian mapping in Regensburg) – Schriftenr Bayer Landesamt f. Umweltschutz 112: 239–252

Muise O, Schäfer N (1992/94) Amphibien in der Stadt Regensburg (Amphibians in Regensburg) – Schriftenr. Bayer. Landesamt f. Umweltschutz 112: 233–238

Schwandorf

Geissner W (1992) Amphibienkartierung im Landkreis Schwandorf (Amphibian mapping in Schwandorf) – Schriftenr Bayer Landesamt f. Umweltschutz 112: 215–221

Spessart

Malkmus R (1974) Die Verbreitung der Amphibien und Reptilien im Hessischen Spessart (The distribution of amphibians and reptiles in Spessart) – Beitr Naturkd Osthessen 9/10: 113–128

Stuttgart

Mertens R (1960) Lurche und Kriechtiere - Stuttgart (Amphibians and reptiles – Stuttgart)

Mertens R (1972) Kriechtiere und Lurche - Stuttgart (Reptiles and Amphibians – Stuttgart)

Weißenburg-Gunzenhausen

Heimbucher D (1992) Amphibienkartierung im Landkreis Weißenburg-Gunzenhausen (Amphibian mapping in Weißenburg-Gunzenhausen) – Schriftenr Bayer Landesamt f. Umweltschutz 112: 123–125

Witten

Sell G, Sell M (1977) Amphibien im Raum Witten (Ruhr) (Amphibians in Witten (Ruhr)) Jb Ver Orts-u. Heimatkd Witten 75: 81–114

Wuppertal

Schall O, Weber G, Pastors J, Gretzke R (1985) Die Amphibien in Wuppertal (The amphibians in Wuppertal) – Bestand, Gefährdung, Schutz – Jahresber Naturwiss Ver Wuppertal 38: 87–107

Italy

Liguria

AAVv (1994) Atlante degli e rettili della Liguria Regione (Atlas of reptiles and the Liguria Region). Liguria ed. Genova

Roma (Rome)

Bologna MA, Capula M, Carpaneto GM, Cignini B, Marangoni M, Venchi A, Zapparoli M (2003) Anfibi e Rettili a Roma – Atlante e guide della specie presenti in citta (Amphibians and Reptiles in Rome – Atlas and Guide to the Species of the city). Comune de Roma, Assessorato Ambienta, Assessorato Cultura Stilgrafica srl, Roma

Poland

Warszawa (Warsaw)

Mazgajska J (1998) Inventory of Amphibians in Warsaw in 1992-1994. In: Barczak T, Indykkiewicz P (eds) (1998). Urban Fauna Wydawnictwo ATR Bydgoszcz (Poland): 227–236

Scotland

Edinburgh

Evans W (1894) On the reptiles and batrachians of the Edinburgh district. Proc Phys Soc Edinburgh 12: 490–526

Switzerland

Bern (*Berne*)

Grossembacher K (1974) Die Amphibien de Umbebung Berns (The Amphbians of Bern). Mit Naturforsch Ges Bern 31: 3–24

Aves (Birds)

Photo: John G. Kelcey

Austria

Linz

Weissmair W, Rubenser H, Brader M, Schauberger R (2001) Linzer Brutvogelatlas (Bird Atlas of Linz) Magistrat der Landeshauptstadt Linz

Wien (*Vienna*)

Aschenrenner L (1974) Liste der Vogel in Raum von Wien (List of Birds in Vienna). Naturgeschichte Wiens, Wien-Munchen 4: 531–536

Boeck F (1993) Die Voegel Wien (Birds of Vienna) Schriffen d.Vereins z. Verbreitung naturwissenschafterlicher Kentnisse in Wien Band 132, Wien

Belgium

Brabant

Herroelen P, de Fraine R (1975) Inventaris van de vogels van Brabant 1900-1974. (Inventory of the birds of Brabant) Ed De Wielewaal-Afdeling, Brussels

Bruxelles (*Brussels*)

Coopman L (1921) Les oiseaux bruxellois (The birds of Brussels). Les Naturalistes belges 2: 103–105

Rabosee D (1989) Atlas des oiseaux nicheurs de Bruxelles (Breeding Bird Atlas of Brussels). Premiers resultants: 1989. Aves 26: 169–182 Leige

Rabosee D (ed) (1995) Atlas des oiseaux nicheurs de Bruxelles 1989-1991 (Breeding Bird Atlas of Brussels). Societ. d'Etudes Ornithol Aves, Leige

Rabosee D, de Wavrin H, Tricot J, van der Elst D (1995) Atlas des Oiseaux nicheurs de Bruxelles 1989-1991 (Atlas of Breeding Birds in Brussels 1989-1991). Soc d'Etudes Ornithol Aves, Liege

Jacob JP, Rabosee D (1995) The Birds of Brussels. Bird Census News 8: 8–21

Weiserbs A, Jacob JP (2007) Oiseaux nicheurs dr Bruxelles 2000–2004 (Breeding Birds of Brussels 2000-2004). Soc d'Etudes Ornithol Aves Liege

Bulgaria

Sofia

Jankov P (1992) Atlas of the Breeding Birds of Sofia. Bird Census News 5: 1–40

Note: In other references elsewhere the author's name is also spelled Iankov or Yankov.

Czech Republic

Pardubice

Vranova S, Lembrek V, Hampl R (2007) Ptaci Pardubic (Pardubice Birds). Ceska Spol Ornitol Pardubice

Praha (*Prague*)

Fric A (1866) Ptactvo města Prahy (Birds of Prague)

Fric A (1912) Ptactvo města Praha. (Birds of Prague) Ziva 13: 96–113

Rozun O (1935) Ptaci v Praze a okoli (Birds in and around Prague)

Wahl V (1944) Prazske ptactvo (Prague Birds)

Wahl V (1945) Prazske ptactvo, 2.vyd. (Prague Birds, 2nd edn) Ceska graficka unie

Fuchs R, Skopek J, Reznicek J (1990) Breeding Bird Distribution Mapping in the Territory of Greater Prague. In: St'ástny K, Bejcek V (eds) Bird Census and Atlas Studies. Proceedings of the XI International Conference on Bird Census and Atlas Work, Prague

Fuchs R, Skopek J, Formanek J, Exnerova A (2002) Atlas hnizdniho rozsireni ptaku Prahy (Atlas of Breeding Birds of Prague). Ceska Spol Ornithol Praha

England

Banbury

Aplin FC, Aplin OV (1882) A List of the Birds of the Banbury District. Banbury Natural History Society, Banbury

Davies G (ed) (1962) The Birds of the Banbury Area. Banbury Ornithological Society, Banbury

Easterbrook TG (1983) Birds of the Banbury Area 1971-81. Banbury Ornithological Society, Banbury

Birmingham

Chase RW (1886) Birds of the Birmingham District. British Association Handbook of Birmingham, Birmingham

Bristol

Bland RL, Taylor SM (1991) The Birdlife of Bristol. In: Frey AE (ed) Bristol's Urban Ecology. Proceedings Bristol Naturalists' Society 48: 63–83

Brighton

Newnham JA (ed) (1988) The Birds of Shoreham and the surrounding area including Brighton, Steyning and Worthing. Shoreham and District Ornithological Society, Shoreham

Bury (Greater Manchester)

Davenport R (1872) Birds of Bury and District

Horan MB (1949) Birds and Mammals of Bury

Calderdale

Murgatroyd F (1986) Birds in Calderdale. Halifax Scientific Society

Murgatroyd F (1987) Birds in Calderdale. Halifax Scientific Society

Murgatroyd F (1989) Birds in Calderdale. Halifax Scientific Society

Chester

Williams TS (1971) Notes on the Bird Life of Chester

CADOS Committee (1984) The Breeding Birds of the Chester Area—results of a tetrad survey

Chorleywood

Drew R, Mitchell J (undated; contents indicate publication after 1996) The Birds of the Rickmansworth and Chorleywood Area

Crewkerne

Parsons AJ (1975) The Breeding Birds of the Parish of Crewkerne. RSPB 55–56: 60

Dartford

Davis WJ (1904) The Birds of the Dartford District. Dartford, Kent

Eastbourne

Arnold EC (1936) Birds of Eastbourne. Strange, Eastbourne

Formby

Wrigley J (1892) The Birds of Formby

Heywood (now in Greater Manchester)

Whittaker I (1935) Birds of a Lancashire Cotton Town—Heywood. Note: the date is also given as 1932.

Leeds

Fuller R, Morris P, Singelton P (1994) Atlas of Breeding Birds in the Leeds area. Leeds Birdwatchers Club, Leeds

Letchworth

Jenkins AR (1958) The Birds of the Letchworth Region. Letchworth Naturalists' Society, Letchworth

Liverpool

Byerley I (1856) The Fauna of Liverpool

Hardy E (1941) The Birds of the Liverpool Area. T. Buncle, Arbroath

London

Hibberd S (1865) London Birds. Intellectual Observer vii: 167–175. London

Power FD (1865) A List of Birds noticed in London during 1863-1864. Zoologist xxiii: 9727

Adams AL (1875) Birds of London Field, London; January 16 and 23

Hamilton E (1879) The Birds of London, Past and Present, Residents and Casuals. Zoologist Vol. III, pp 273–291

Pigott DJ (1884) London Birds and London Insects' 8vo London. Revised 1892 and 1902

Hamilton E (1889) The Wild Birds of London. Murray's Magazine: May 1889

Swann HK (1893) The Birds of London. Swann Sonnenschein

Anon (1898) The Birds of London. Edinburgh Review: London, January 1898

Hudson WH (1898) Birds in London. Longmans, Green and Co. London. Reprinted 1924 by JM Dent, London and c.1969 by David and Charles Reprints, Devon

Dixon C (1909) The Bird Life of London. Heinemann, London

Harding JR (1922) Birds of London

Fitter RSR, Parrinder ER (1944) A Check-list of the Birds of the London Area. London Bird Report 1943: 20–28

Fitter RSR (1949) London's Birds. Collins, London

Homes RC, Acland CM, Ashby CB, Collenette CL, Fitter RSR, Parrinder ER, Richards BA (1957) The Birds of the London Area since 1900. W. Collins, London

Note: in some references the publisher is also given as London Natural History Society, (1957).

Homes RC (ed) (1964) The Birds of the London Area. Rupert Hart-Davis, London

Note: this reference is mentioned as a "new, revised edition" of the 1957 reference given above.

Montier DJ (1977) Atlas of Breeding Birds of the London area. London Natural History Society, B.T. Batsford, London

Nicholson EM (1995) Bird-watching in London. London Natural History Society

Hewlett J (2002) The Breeding Birds of the London area. London Natural History Society, London

Self A (2014) The Birds of London. Bloomsbury, London

Manchester (Greater Manchester)

Holland PK, Spence IM, Sutton JT (1984) The Breeding Birds in Greater Manchester. Manchester Ornithological Society

Smith AJ (ed) (1996) Birds of Greater Manchester. Twentieth Greater Manchester Bird Report 1995. The Greater Manchester Birds Club

Newbury

Palmer, MH (1886) The Birds of Newbury. Newbury Weekly News 29 May to 10 June

Summers-Smith JD (1950) A History of the Birds of Newbury District. Transactions, Newbury District Field Club 9: 59–86

Summers-Smith JD (1954) A History of the Birds of Newbury District 1951-53. Transactions, Newbury District Field Club 10: 26–31; 59–86

Lewis LR (1958) Birds of the Newbury District 1954-58. Transactions, Newbury District Field Club 10: 14–30

Oldham (Greater Manchester)

Stubbs F J (1905 and 1910) Birds of the Oldham District

Oxford

Bayliss MJ (1982) Birds of Oxford City. Oxford Ornithological Society Report 1982: 56–66

Plymouth

Curry D (undated; possibly about 1988) Birds of Plymouth
Tucker V (1995) Birds of Plymouth. Devon Bird Watching and Preservation Society, Plymouth

Reading

Balch CC (1951) Provisional List of the Birds of Reading. Reading Ornithological Club, Reading

Rickmansworth

See Chorleywood

Rochdale (Greater Manchester)

Williamson F, Lord W (1913) Birds of Rochdale District

Sandbach

Sibson RB (1945 and 1946) Notes on the Birds of Sandbach, SE. Cheshire. Northwestern Naturalist

Sheffield

Smith AHV (1974) Birds of the Sheffield Area. Sorby Record Special Series
Smith AHV (1985) Birds. In: Whiteley D (ed) The Natural History of the Sheffield Area. Sorby Natural History Society, Sheffield, pp 105–121

Shoreham

Newnham JA (ed) (1988) The Birds of Shoreham and the surrounding area including Brighton, Steyning and Worthing. Shoreham and District Ornithological Society, Shoreham

Southport

Wagstaffe R (1935) Birds of Southport

Spalding

Brogden THJ (1900) Birds of Spalding and the South Lincolnshire Fenland 1886-1896. The Naturalist 1900, pp 17–32

Staines

Kerr GW (1906) Birds of the District of Staines. Zoologist 64: 179–184; 230–234; 307–310 and 386–389
Kerr GW (1908) Birds of the District of Staines. Zoologist 66: 137–143

Steyning

Newnham JA (ed) (1988) The Birds of Shoreham and the surrounding area including Brighton, Steyning and Worthing. Shoreham and District Ornithological Society, Shoreham

Stoneyhurst

Howard RJ, Mitchell FS (1888) The Birds of Stoneyhurst

Todmorden

Standsfield A (ed) (1911) Birds of Todmorden

Wilmslow

Bell TH, Samuels LP, Pownall LA (1955) Notes on the Birds of the Urban Area of Wilmslow

Winchester

Campbell B, Pringle JWS, Hasler WJ (1931) The Birds of Winchester. Winchester College Natural History Society Report 1927-31: 22–42

Cheke AS (1962) Birds of Winchester and District. P&G Wells, Winchester

Worcester

Curtler M (1853) (title not known) In: Stanley's Guide to Worcester. Worcester

Worthing

Newnham JA (ed) (1988) The Birds of Shoreham and the surrounding area including Brighton, Steyning and Worthing. Shoreham and District Ornithological Society, Shoreham

France

Douai

Boutroille C (2005) Atlas des oiseaux nicheurs de ville de Douai: bilan provisuar de deux annes de prospection: 2003 et 2004 (Breeding Bird Atlas of the city of Douai; interim evaluation of two years of prospecting 2003 and 2004). Le Heron 38

Paris

Paquet R (1874) Ornithologie parisienne (Paris Ornithology) Parus

Legendre M (1928) Les oiseaux de Paris (The Birds of Paris). Revue Francaise d'orthithologie XII: 92–111

Malher F, Lesaffre G, Zucca M, Coarmeur J (2010) Oiseaux nicheurs de Paris. Un atlas urbain (Breeding birds in Paris. An urban atlas). Delachaux et Niestle, Paris

Germany

Berlin

Bruch A, Elvers E, Pohl CH, Westphal, Witt K (1978) Die Vogel in Berlin (West) (The Birds of West Berlin). Eine ubersicht Orn Ber f Berlin 3 Sonderheft 286 S

Ornithol. AG Berlin (West)(1984) Brutvogelatlas Berlin (West)(Breeding Bird Atlas of Berlin (West) Ornitho). Ber F-Berlin (west) 9, Sonderheft, 384 S (Note: also referred to as Witt K (1984))

Degen G, Otto W (1988) Atlas der Brutvogel von Berlin (Atlas of Breeding Birds of Berlin) Natureschutzarbeit Berlin Brandenburg, Beiheft 8, Berlin u. Potsdam

Ornithol. AG Berlin (West) (eds) (1990) Die Vögel in Berlin (West) Eine Übersicht. Ergänzungsbericht 1976-1989 (The Birds of Berlin (West) – an overview. Supplementary Report 1976-1989) Ornithol. Ber f Berlin (West) 15, Sonderheft

Otto W, Witt K (2002) Verbreitung und Bestand berliner Brutvögel. Berliner ornithol. (Distribution and status of breeding birds in Berlin). Bericht. Band 12, Sonderheft, Berlin. Note: this is the first book of the breeding birds of the whole city.

Bielefeld

Laske V, Nottmeyer-Lindon K, Conrads K (1991) Die Voegel Bielefelds (Birds of Bielefeld). Ilex-Buecher, Natur B. 2, Bielefeld

Bonn

Rheinwald G, Wink M, Joachim H-E (1984) Die Vögel im Grossraum Bonn. Beitraege z. Avifauna des Rheinlandes, (Birds of the Bonn Area). Heft 22/23, Duesseldorf

Rheinwald G, Kneitz S (2002) Die Vögel zwischen Sieg, Ahr und Erft. Ginster-Verlag, St. Katharinen. According to the authors the title “has no English translation”.

Chemnitz

Flöter E, Saemann D, Börner J (2006) Brutvogelatlas der Stadt Chemnitz (Atlas of the Breeding Birds of Chemnitz). Mitteilungrn d. Vereins Saechsischer Ornithol. Band 9, Sonderheft 4

Dortmund

Kretzschmar E, Neugebaur R (2003) Dortmunder Brutvogelatlas (Dortmund Breeding Bird Atlas). NABU-Stadtverband Dortmund

Dusseldorf

Leisten A, Kämpe BSU (2002) Die Vogelwelt der Stadt Düsseldorf: Brutvogelatlas mit avifaunistischen Beitragen. (The bird life of the city of Dusseldorf: Breeding Bird Atlas with avifaunistic contribution). Bologische Station Urdenbacher Kämpe NABU-Stadtverband Düsseldorf

Emden

Retting K (2007) Brutvogelatlas Stadt Emden (Emden Breeding Bird Atlas) Emden, Stadt Emden Fachdienst Umwelt

Essen

Przygodda W (1998) Die Voegel von Essen und Muelheim an der Ruhr (Birds of Essen and Mulheim in the Ruhr). Gesel Rheinischer Ornithol heft 29. Dusseldorf

Halle

Gnielka R. Mitarbeit von T. Spretke, H. Tauchnitz, Reuter B (1983) Avifauna von Halle und Umbebung 1 (Birds of Halle and vicinity No. 1). Herausgeber: Rat der Stadt Halle (Saale) Abteilung Umweltschutz und Wasserwirtschaft und Gesellschaft für Natur und Umwelt, Kreisvorstand Halle, im Kulturbund der DDR

Gnielka R. Mitarbeit von Dr. K. Liedel, T. Spretke, Dr. A. Stiefel, H. Tauchnitz, Wischhof W. (1984) Avifauna von Halle und Umbebung 2 (Birds of Halle and vicinity No. 2). Herausgeber: Rat der Stadt Halle (Saale) Abteilung Umweltschutz und Wasserwirtschaft und Gesellschaft für Natur und Umwelt, Kreisvorstand Halle, im Kulturbund der DDR

Halberstadt

Nicolai B, Wadewitz M (2003) Die Brutvögel von Halberstadt. Ergbnisse einer Brutvogelkarteirung 1988 bis 2002 (The breeding bird of Halberstadt; results of a breeding bird mapping 1988-2002). Musuem Heinearum, Halberstadt

Hamburg

Dietrich F (1928) Hamburgs Vogelwelt (Hamburg birdlife) Hamburg

Holzapfel C, Hüppop O, Mulsow R (1984) Die Vogelwelt von Hamburg und Umgebung Band I (Birds of Hamburg and vicinity vol. I). Karl Wachholtz Verlag, Neumünster

Holzapfel C, Hüppop O, Mulsow R (1984) Die Vogelwelt von Hamburg und Umgebung Band II (Birds of Hamburg and vicinity vol. II). Karl Wachholtz Verlag, Neumünster

Garthe S (1996) Die Vögel von Hamburg and Umgebung Band III (Birds of Hamburg and vicinity vol. III). Wacholtz Verlag, Neumünster

Mitschke A, Baumung S (2001) Brutvogel-Atlas Hamburg (Atlas of Breeding Birds of Hamburg vol. 31). Hamburger Avifaunistische Beiträge (hab) Band 31, Hamburg

Mainz

Thomas B (1983) Zur Avifauna von Mainz (The Avifauna of Mainz). Mainzer Naturwissenschaftliches Archiv 3

Melle

Tiemeyer V (1993) Die vogel der Stadt Melle (The birds of Melle) E. Knoth, Melle

Muelheim

Przygodda W (1998) Die Voegel von Essen und Muelheim an der Ruhr (Birds of Essen and Mulheim in the Ruhr). Gesel Rheinischer Ornithol heft 29. Dusseldorf

Oberhausen

Hyla W (1991) Die Vogelwelt von Oberhausen (The birdlife of Oberhausen) Stadt Oberhausen, Amt f. Umweltschutz 206 S. (abb Kt-Sk, Tab, Qu)

Regensburg

Klose A, Leibl F, Vidal A (1986) Avifauna Ratisbonensis. Die Vogelwelt der Stadt Regensburg (The birds of Regensburg). Acta Albertina Ratisbonensia 43

Schlemmer R, Vidal A, Klose A (2013) Die Brutvögel der Stadt Regensburg und ihre Bestandsentwicklung von 1982 bis 2012 (The breeding birds of Regensburg and its population trends from 1982 to 2012). Acta Albertina Ratisbonensia (Sonderheft)

Saarbrücken

Handke K, Petermann P, Ellenberg H, Hayo L (1986). Atlas der Vögel des Saarbrücker Raumes (Bird Atlas of the Saarbrücken area). Universität des Saarlandes, Saarbrücken

Hungary

Debrecen

Bozsko S, Juasz L (1983) Die Vogelwelt der Stadt Debrecen und ihrer Veranderungen in diesen Jahrhundert (The birdlife of Debrecen and its changes in this century) Dery Muzeum Enkonyve

Italy

Bergamo

Cairo E, Facoetti R (2006) Atlante degli uccelli di Bergamo (Atlas of the Birds of Bergamo). Edizioni Junior, Azzano San Paolo

Biella

Bordignon L (1999) Gli uccelli citta di Biella (Atlas of the Breeding Birds of Biella). Commune di Biella ed Biella

Brescia

Ballerio G, Brichetti P (2003) Atlante degli uccelli nidificanti nella citta di Brescia 1994-1998 (Atlas of the Breeding Birds of Brescia). Natura Brescinia 33: 133–167

Cagliari

Mocci Demartis A, Gruppo ICNUSA (1992) Atlanti ornitologici urbani Cagliari (Ornithological atlases of urban Cagliari). Ecologia Urbana 4 (2–3): 22–23

Cossato

Bordignon, L (1997) Atlante degli uccelli nidificanti a Cossato (Atlas of the Breeding Birds of Cossato). Quaderni di educazione ambientale, Commune di Cossato ed, Cossato

Crema (CR)

Mastrorilli M (2002) La communita ornitica nidificante della citta di Crema (CR) nel biennion 2000-2001 (The Community breeding bird population of the city of Crema (CR) in 2000-2001). Rivista Italiana di Ornitologica 72: 261–268

Cremona

Groppali R (1994) Gli degli uccelli nidificanti e svernanti nella citta di Cremona (1990-1993) (Atlas of the Breeding Birds of Cremona). Azienda Energetica Municipalizzata di Cremona e Museo Civico di Storia Naturale di.Cremona, Cremona

Groppali R (2004) Nidificanti e svernanti a Cremona tra 1990-1993 e 2001-2004 (Nesting and wintering in Cremona between 1990 and 1993 and 2001 and 2004). Comune di Cremona, Museo Civico di Storia Naturale, Cremona

Firenze (*Florence*)

Dinetti M, Ascani P (1990) Atlante degli ucelli nidificanti nel commune di Firenze. (Atlas of the Breeding Birds of Florence). Comune di Firenze, Firenze

Dinetti M, Romano S (2002) Atlante degli ucelli nidificanti nel commune di Firenze 1997-1998. (Atlas of the Breeding Birds of Florence 1997-1998). Comune di Firenze, Firenze

Dinetti M (2009) Atlante degli ucelli nidificanti nel commune di Firenze Terza edizione 2007-2008. (Atlas of the Breeding Birds of Florence 3rd edn 2007-2008). LIPU ed., Firenze

Forli

Ceccarelli PP, Gellini S, Casadei M, Ciani C (2006) Atlante degli uccelli nidificanti a Forli (Atlas of the Breeding Birds of Forli). Museo Ornitologica "Ferrante Foschi" ed. Forli

Genova (*Genoa*)

Borgo E, Galli L, Galluppo C, Maranini N, Spano S (2005) Atlante ornitologico della citta di Genova (Atlas of the Birds of Genoa). Bollettino dei Musei e degli Instituti Biologici dell'Universita di Genova, 69–70: 1–317

Grosseto

Giovacchini P (2001) Atlante degli uccelli nidificanti a Grosseto (Atlas of the Breeding Birds of Grosseto). Supplemento 17 degli Atti del Museo di Storia Naturale della Maremma

Giovacchini P (2011) Nuovo Atlante degli uccelli nidificanti a Grosseto (2009-2010) (New Atlas of the Breeding Birds of Grosseto). Commune di Grosseto Assessorato all/Ambiente, Tipografia Myckpress, Fornacette (PI)

La Spezia

Biagioni M, Coppo S, Dinetti M, Rossi E (1996) La Conservazione della biodiversta ne commue della Spezia. (Biodiversity conservation of La Spezia). Commue della Spezia

Dinetti M (1996) La conservazione della biodiversita nel commune di La Spezia. (Biodiversity conservation of La Spezia) Commune della Spezia ed, La Spezia

Latium

Alexander HG (1927) The Birds of Latium, Italy. The Ibis 69: 245–283

Lazio

Boano A, Brunelli M, Bulgarini F, Montemaggiori A, Sarrocco S, Visentin M (1995) Atlante degli uccelli nidificanti nel Lazio (Atlas of breeding birds in Lazio) Alula, volume speciale (1–2): 1–224

Biondi M, Guerrieri G, Pietreli L (1999) Atlante degli uccelli presenti in inverno lungo la fascia costiera del Lazio (1992-1995) (Atlas of birds present in winter along the coast of Lazio (1992-1995)). Alula, VI: 3–124

Brunelli M, Sarrocco S, Corbi F, Sorace A, Boano A, de Felici S, Guerrieri G, Meschini A, Roma S (2011) Nuovo Atlante degli uccelli nidificanti nel Lazio (New Atlas of breeding birds in Lazio) Agenzia Regionale per I Parchi ed. Roma

Lido di Ostia

Demartini L, Sorace A, Cecere JG, Savo E, Polinori A, Cignini B, Zapparoli M (1996) Atlante degli uccelli nidificanti centro urbano di Lido di Ostia (Atlas of breeding birds in the urban centre of Lido di Ostia) Associazione Centro Habitat Mediterraneo ed. Roma

Liguaria (*Liguria*)

AAVv (1989) Atlante degli uccelli nidificanti in Liguria (Atlas of breeding birds in Liguaria) Regione Liguria ed, Genova

Livorno

Dinetti M (1994) Atlante degli uccelli nidificanti di Livorno (Atlas of the Breeding Birds of Livorno). Comune di Livorno ed, Quaderni dell'ambiente, 5. Livorno

Dinetti M, Ascani P, Franceschi A, Tiengo M, Arcamone E (2013) Atlante degli ucelli nidificanti a Livorno 2006–2013 (Atlas of breeding birds in Livorno 2006–2013) Felici Editore, Pisa

Milano (Milan)

Nova M (2002) Da Moltoni al 2000: le conoscenze sugli uccelli nidificante a Milano (From Moltoni to 2000: the knowledge of birds nesting in Milan). Rivista Italiana di Ornitologia, 72: 127–149

Bonazzi P, Buvoli L, Belardi M, Brambilla M, Celada C, Favaron M, Gottardi G, Nova M, Rubolini D, Fornasari L (2005) Il progetto AVIUM (Atlante virtuale degli uccelli di Milano) (The project AVIUM (Virtual Atlas of Birds of Milan)) Ecologia Urbana 17: 13–16

Monti Lepini

Corsetti L (1989) Atlante ornitologico dei Monti Lepini (Ornithological Atlas of Lepinis) Ypothekai

Napoli (Naples)

Fraissinet M (1984) L'avifauna di Napoli (Birds of Naples). Avocetta: 8–19

Fraissinet M (1995) Atlante degli ucelli nidificanti e svernanti nella citta di Napoli (Atlas of the Breeding and wintering Birds of Naples). Assocciazione Studi Ornithol. Italia Meridionale Monografia n. 4, Napoli

Fraissinet M (2006) Atlante degli uccelli nidificanti e svernanti nella citta di Napoli 2001-2005 (Atlas of the Breeding and wintering Birds of Naples 2001-2005). Associazione Studi Ornithol. Italia Meridionale, Monografia n. 7, Napoli

Padova

Bottazzo S, Giacomini G (2010) Atlante degli uccelli nidificanti a Padova (Atlas of breeding birds of Padova) LIPU, sezione di Padova, Grafiche Erretre, Pernumia (Padova)

Palermo

Lo Valvo M, La Mantia T, Massa B (1985) Bird Population of Palermo's urban and suburban areas. Boll. Zool. 52: 347–354. Note: This may have been published in Italian.

Pavia

Bernini F, Dinetti M, Garribaldi A, Matessi G, Rognoni G (1998) Atlante degli uccelli nidificanti a Pavia (Atlas of the Breeding Birds of Pavia) Commune di Pavia e LIPU, Pavia

Piacenza

Ambrogio A, Figoli G, Ziotti L (eds) (2001) Atlante degli uccelli nidificanti nel piacentino (Atlas of breeding birds in Piacenza) LIPU, Sez.di Piacenza ed

Pisa

Dinetti M (1998) Atlante ornitologico dell'area urbana del commune di Pisa 1997-1998 (Atlas of the Breeding Birds of Pisa 1997-1998). LIPU e Comune di Pisa

Dinetti M (2003) Atlante degli uccelli nidificanti nell'area urbana di Pisa (Atlas of the Breeding Birds of Pisa 1997-1998). Comune di Pisa e LIPU. WEB site of the Comune di Pisa

Reggio Emilia

Gustin M (2002) Atlante degli uccelli nidificanti a Reggio Emilia (Atlas of the Breeding Birds of Emilia) Comune di Reggio Emilia ed

Roma (*Rome*)

Cignini B, Zapparoli M (eds) (1996) Atlante degli uccelli nidificanti a Roma (Atlas of the Breeding Birds of Rome). Fratelli Palombi Editori, Roma

Cardillo A, Cento, M, Fraticelli F (2011) Atlante degli uccelli in inverno a Roma: 2007-2011 (Atlas of birds in winter in Rome: 2007-2011). Abstract X Conv Ital Ornitol *Cervia*: 111–112

San Dona di Piave, Venezia (*Venice*)

Marcolin C, Zanetti M (1999) Uccelli in citta. Atlante degli uccelli nidificanti nell'area urbana di San Dona e Musile di Piave (Atlas of the Breeding Birds of San Dona and Venice). Edizioni Associazone Naturalistics Sandonatese, Noventa di Piave (Venice)

Nardo A (2003) Atlante degli uccelli nidificanti e svernanti a San Dona di Piave (Atlas of the Breeding Birds of San Dona). Comune di San Dona di Piave e LIPU eds

Torino (*Turin*)

Isaaia A, Dotti L (1989) Avifauna del commune di Torino (Birds of Turin). Riv. Piem. St. Nat. 10: 253–279

Maffei G, Pulcher C, Rolando A, Carisio L (2001) L'avifauna della citta di Torino: analisi ecologica e faunistica (The avifauna of the city of Turin: ecological analysis and fauna). Monografie XXXI, Museo Regionale di Scienze Naturali, Torino

Trento

LIPU (1998) Atlante degli uccelli nidificante nel commune di Trento (Atlas of birds nesting in the commune of Trento) Natura alpine 48: 1–207

Treviso

Nardo A, Mezzavilla F (2011) Progetto Atlante degli uccelli nidificante e svernanti nel commune di Treviso (Project Atlas of birds nesting and wintering in the commune of Treviso) Abstract XVI Convegno Italiano di Ornitologia Cervia 128–129

Udine

Parodi R (2008) Avifauna del Comune di Udine (Avifauna of the Municipality of Udine) Publicazione n.51 – Edizioni del Museo Friulano di St. Naturale. Commune di Udine

Umbria

Magrini M, Gambaro C (1997) Atlante ornitologico dell'Umbria (Ornithological atlas of Umbria). Regione dell'Umbria ed. Citta di Castello

Valsesia

Bordignon L (1993) Gli uccelli della Valsesia (The birds of Valsesia) C.A.I. Varallo ed

Varese

Vigano A (1996) Il Progetto Atlante degli uccelli nidificante nel Comune di Varese (The Project Atlas of birds nesting in the town of Varese) Proceedings Conv Naz. "L'Avifauna degli ecosistemi di origine antropica." Monografia n.5 dell'ASOIM, Napoli, pp 46–47

Verano Cusio Ossola

Biona R, Bordignon L (eds.) (2006) Atlante degli uccelli nidificanti del Verano Cusio Ossola (Atlas of breeding birds of Verano Cusio Ossola) Quaderni di Natura e Paesaggio del Verbano Cusio Ossola, Numero 6

Viterbo

Cignini B, Meschini A, Zapparoli M (1994) Progetto Atlante degli uccelli nidificanti a Viterbo: prima stagione di rilevamento (1991) (Project Atlas of breeding birds in Viterbo: the first season of detection (1991)) In: Baldaccini NE Mingozzi T, Violani C (eds) Atti del VI Convegno Italiano di Ornitologia, (Proceedings of the VI Congress of Italian Ornithology) Museo Regionale Scienze Naturali, Turino, pp 532–534

Lithuania

Vilnius

Anon (1990) Fauna Lietuvos: paukščiai Vilnius miestas sostinė Makstas (Fauna of Lithuania: birds in Vilnius capital city of Muskat)

Idzelis R (1993) Avifauna of Vilnius Act Orthologica, Litianica 7–8: 86–92

Note: not known whether published in English or Lithuanian

Poland

Częstochowa

Czyż S (2008) Atlas ptaków lęgowych Częstochowy 2003–2007. Atlas of Breeding Birds in Czestochowa). Wyd. S. Czyż, Częstochowa. Stanislav Czyz, Czestochowa 2008

Gorzow

Jermaczek A, Czwałga T, Jermaczek D, Nowak A, Stanko R, Scheinder G, Zeglinski G (1992) Ptaki legowe miasta Gorzowa w roku 1989 (Breeding birds of Gorzów in 1989) Lubuski Przeglad Przyrodniczy 3: 41–69

Jaslo

Stoj M, Dyczowksi J (2002) Ptaki Jasla liczebnose, rozmieszczenie I ochrona (Birds of Jaslo, Numbers, Distribution and Conservation). Bogucki wyd. Naukowe, Poznan

Leszno

Kuzniak S (1996) Atlas ptakow legowych Leszna w latach 1990-1993 (Atlas of Breeding Birds of Leszno in 1990-1993). Prace Zakl Biol Ekol Ptrakow UAM 6 Poznan

Lublin

Riabinin S (1959) Ptaki Lublina w latach 1951/56 (The Birds of Lublin 1951-1956) Ochr Przyr 26: 419–449

Biadun W (2004) Ptaki Lublina (Birds of Lublin) Akad. Medyczna w Lubliniie, Lublin

Łodź

Janiszewski T, Wojciechowski Z, Markowski J (2009) Atlas ptakow legowych Łodzi (Atlas of Breeding Birds of Łodź). Wyd Uniw Lodzkiego Łodź

Olsztyn

Okulewicz J (1971) Birds of Olsztny and its vicinity. Acta orn 13: 127–172

Nowakowski J, Dulisz B, Lewandowski K (2006) Ptaki Olsztyna (Birds of Olsztyn) Prac Wydawn "Elset", Olsztyn

Poznan

Ptasyk J (2003) Ptaki Poznania – stan jakosciowy I ilosciowy oraz jeo w latach 1850-2000 (Birds of Poznan—status and changes between 1850 and 2000). Wyd Naukowe UAM, Poznan

Stupsk

Gorski W (1982) Ptaki lęgowe Stupsk i jej obszarach podmiejskich (Breeding Birds of Stupsk and its suburban areas) Acta zool Cracov 26: 31–93

Sulechowa

Czwalga T (1992) Awifauna legowa miasta Sulechowa w latach 1988-89 (The breeding birds in Sulechów in 1988-89) Lubuski Przeglad Przyrodniczy 3: 13–69

Świebodzina

Jermaczek A, Jermaczek D, Filipczak K (1990) Ptaki legowe miasta Swiebodzina w latach 1988-1989 (Breeding birds of Świebodzin in 1988-1989) Lubuski Przeglad Przyrodniczy 1: 3–33

Warszawa (Warsaw)

Luniak M, Kalbarczyk W, Pawlowski W (1964) Ptaki Warszawy (Birds of Warsaw). Acta Ornithol Warszawa 8: 175–285. Note: some other references give the page numbers as 176–286.

Luniak M, Kalbarczykw W, Pawloski W (1964) Ptaki Warszawy (Birds of Warsaw) Acta Ornithologica 8 (6): 175–285

Luniak M (1965) Ptaki naszej stolicy (Birds of Our Capital). Przyr Pol. 9: 5

Luniak M (1974) O Ptakach Warszawy (On the Birds of Warsaw). Nasza Ksiegarnia, Warszawa

Luniak M (1996) In: English - Inventory of the avifauna of Warsaw – Species Composition, Abundance and Habitat Distribution. Acta Ornithologica 31: 67–80

Luniak M, Kozlowski P, Nowicki W, Plit J (2001) Ptaki Warszawy 1962-2000 (Birds of Warsaw 1962-2000). IGiPZ PAN, Warszawa

Nowicki W (2001) Ptaki śródmieścia Warszawy (Birds of Warsaw). Muz. i Inst. Zoologii PAN, Warszawa

Portugal

Lisboa (Lisbon)

Elias G, Peino L (1997) Guia das Aves de Lisboa (Guide to the Birds of Lisbon). Câmara Minicipal de Lesboa, Pebouro do Ambiente e Espaeos Verdes

Costa H, Oliveira M (2000) Lisboa Aves (Lisbon Birds). Câmara Minicipal de Lesboa e Instituto de Conservacâo da Natureza

Russian Federation

Note: it is assumed that most if not all of the original references are in Cyrillic.

Москва (*Moscow*)

Ilichev VD, Butiev VT, Konstantinov VM (1987) Птицы Москвы и его окрестности (Birds of Moscow and its vicinity)

Kalyakin MV, Volzit OV (2006) Atlas of Birds of Moscow and the Moscow Region. Pensoft, Sofia & Moscow

Kalyakin MV, Voltzit OV, Groot Koerkamp G (2014) Atlas ptici goroda Moskvy (Atlas of the Birds of Moscow city) Fiton XXI, Moscow

St. Petersburgh (formerly Leningrad)

Anon (1986) Ekologiceskiye i faunisticeskiye issledovaniya ptic, Trudy Zoologiceskogo Instituta Akademii Nauk SSSR, vol. 147, pp 89–121 (Review of avifauna of Leningrad in vol. Ecological and faunistic studies on birds) in the series Works of the Zoological Institute of the Academy of Sciences USSR)

Khrabryi VM (1991) Птицы Санкт-Петербурга; ДАН СССР, Тр Зоол. Текущего месяца Том 236, Санкт-Петербург (The Birds of St. Petersburg) Akad Nauk SSSR, Trudi ZIN Russian Academy of Sciences, Tom 236

Храбрый ВМ (1992) Орнитофаунистическая карта (Распространение гнездящихся птиц). В кн.: Экологический атлас Санкт-Петербурга. СПб. (Khrabryi V (1992) Ornithology Map (Distribution of Breeding Birds). In: Ecological Atlas of St. Petersburg-St. Petersburg)

Khrabryi VM (2002) Dynamics of species composition and numbers of breeding birds in different biotopes of Saint-Petersburg during 24 years. Proceed. Zool. Inst. Russ Acad. Sci.

Voronezh

Numerov AD, Vengerov PD, Kiselev OG, Boriskin DA, Vetrov EV, Kireev AV, Smirnov SV, Sokolov AYu, Uspensky KV, Shilov KA, Yakovlev YuV (2013) (Atlas of breeding birds in the city of Voronezh). Izd. Naucnaya Kniga, Voronezh

Note: a similar publication appears under the following reference:

Sokolov AYu, Uspiensky KV, Shilov KA, Yakovlev YuV (2013) (Atlas of breeding birds in the city of Voronezh). Izd. Naucnaya Kniga, Voronezh

Serbia

Belgrade

Vasic V (1967-1968) The Birds of Belgrade. Larus 21–22: 130–136

Spain

Valencia

Murgui E (1998) First results of the wintering bird atlas of Valencia city. Biologia e Conservazione della Fauna 102: 155–162

Switzerland

Basle/Basel

Blattner M, Kestenholz M (1999) Die Brutvoegel beider Basel (Breeding Birds of Basel). Mitt. Naturforschenden Gesel. Basel, vol. 4, Liestal

The Netherlands

Alkmaar

Smit H, Roobeek CF, Damm T (2005) *De Broedvogels van Alkmaar in 2001–2004. (The breeding birds of Alkmaar 2001–2004)* Heerhugowaard: Vogelwerkgroep Alkmaar e.o.

Amsterdam

Melchers M, Daalder R (1996) Sijsjes en Drijfsijsjes de vogerls van Amsterdam (Siskins and Floating Ice. The birds of Amsterdam). Schuijt & Co, Haarlem

Breda

Van Iersel P, Van Iersel P, Van Der Sanden A (2009) *Vogels Houden van Breda.* Breda: (Birds love Breda) NPN Media B.V.

Delft

Mostert K (1995) De broedvogels van het stedelijk gebied van Delft in 1990 : met een overzicht van inventarisaties uit de drie voorgaande decaden en de jaren 1991 tot en met 1994 (The breeding birds of the urban area of Delft in 1990 with a review of inventories from the previous three decades and the years 1991 to 1994). Vogelwerkgroep Delft

Eindhoven

Marechal P, Veenhuizen, W (1997) *Vogels in het Stedelijke Millieu: Inventarisatie in Eindhoven.* (Birds in the Urban Environment: Inventory of Birds of Eindhoven) Utrecht: Wetensch Med. KNNV.

Niuwegein

Abel G et al. (2009) *Broedvogels in Nieuwegein: Waar, hoeveel en trends.* (Breeding birds in Niuwegein: number, frequency in trends) Utrecht: Vogelwacht Utrecht

Ukraine

Publication language not known

Lutsk

Khymyn M (1993) The Atlas of Wintering Birds in Lutsk District (1988/89-1991/92). Bird World, Lutsk

Lvov

Bokotev A (date not known; post 1990) General characteristics of the avifauna of Lvov, Ukraine 1990

Mammalia (Mammals)

Photo: Jaroslav Červeny

Czech Republic

Brno

Pelikan J, Zejda J, Homolka M (1983) Mammals in the urban agglomeration of Brno. Prirodoved. Pr Ustavu Cesk Akad Ved Brno 17 Nr 9 S.: 1–49

Praha (Prague)

Hanak V (1975) Prazsti savci (Mammals of Prague). Ziva 23: 235–237
Hanak V (1976) Prazsti savci (Mammals of Prague). Ziva 24: 35–37 and 71–73

England

Bristol

Charbonnier HJ, Lloyd MCC (1898) The Mammals of the Bristol District. In: Rogers BMH (ed) Handbook to the Bristol Neighbourhood. Brit Assoc Bristol 1898. J Wright, pp 200–203
Rudge CR, Charbonner HJ (1909) The Mammals of the Bristol District. Proc Bristol Naturalists' Soc 2, pp 55–60

Bury

Horan MB (1949) Birds and Mammals of Bury

Derby

Middleton JF (1969) Mammals of the Derby Area. Derby Junior Naturalists, Derbyshire

Halifax

Morley I (1965) Vertebrate Fauna of the Halifax Parish. Halifax Scientific Society

London

Stubbs FJ (1917) The Mammals of the London District. School Nature Study 1917

Fitter RSR (1949) A Check-list of the Mammals, Reptiles and Amphibian of the London Area 1900-1949. London Naturalist 28: 98–115

Fitter RSR (1960) Further Records of Mammals, Reptiles and Amphibians in the London Area. London Naturalist 39: 18–21

Teagle WG (1963) Mammals in the London Area: a report for 1960 with some records for the years 1957-1959. London Naturalist 42: 42–56

Teagle WG (1964) Mammals in the London Area 1961. London Naturalist 43: 119–135

Teagle WG (1965) Mammals in the London Area 1962. London Naturalist 44: 43–57

Burton John A (1966) Notes of the Mammals of the London Area during 1963 and 1964. London Naturalist 45: 29–34

Burton John A (1967) Notes on the Mammals in the London Area for 1965. London Naturalist 46: 40–43

Beames IR (1968) Mammals in the London Area 1966. London Naturalist 47: 25–37

Beames IR (1969) Mammals in the London Area 1967. London Naturalist 58: 40–47

Gold KAJ (1975) Mammals in the London Area 1974. London Naturalist 54: 66–68

Gold KAJ (1976) Mammals in the London Area 1975. London Naturalist 55: 58–61

Hall DG (1978) Mammals in the London Area 1976. London Naturalist 57: 92–95

Hall DG (1978) Mammals in the London Area 1977. London Naturalist 57: 96–97

Cotton J (1979) Mammals in the London Area 1978. London Naturalist 58: 71–73

Cotton J (1980) Mammals in the London Area 1979. London Naturalist 59: 84–86

Herbert C (1996) A check-list of the mammals of the London area 1900-1994. London Naturalist 75: 91–94

Scarborough

Rimington FC (1956) Mammals. In: Walsh GB, Rimington FC (eds) The Natural History of Scarborough District II. Scarborough Field Naturalists' Society, pp 408–416

Sheffield

Clegg TM (1963) The Mammals of the Sheffield Area. Naturalist 103–105

Whiteley D, Whiteley SS (1976) Mammal Fauna of Sheffield: Part 1 Sorby Record 4–31

Clinging V, Whiteley D (1980) Mammals of the Sheffield Area. Sorby Record, Special Series 3: 1–48. Sorby Natural History Soc Sheffield City Museum

Note: references also given as Clinging V, Whiteley D (1980) Mammals. In: Whiteley D (ed) The Natural History of the Sheffield Area. Sorby Record Special Series 3. Sorby Natural History Society, pp 1–48

York

Grabham O (1906) Mammals. In: Auden GA (ed) Historical and Scientific Survey of York and District. British Association for the Advancement of Science

Scotland

Glasgow

Scott-Elliot GF, Laurie M, Murdoch JB (1901) Fauna, Flora and Geology of the Clyde Area Glasgow. Local Committee for the meeting of the British Association

Pisces (Fish)

Photo: Lubomir Hlašek

England

London

Wheeler AC (1958) Fishes of the London Area. London. Naturalist 37: 80–101

Marlborough D (1962) A supplement to the "Fishes of the London Area." London. Naturalist 42: 62–70

Marlborough D (1965) London Fish since 1962. London Naturalist 44: 70–73

Marlborough D (1972) London Fishes to 1971. London Naturalist 50: 63–78

Sheffield

Mander PB, Riley TH, Whiteley D (1976) Freshwater Fishes of the Sheffield Area. Sheffield City Museums

York

Fortune R (1906) Pisces. In: Auden GA (ed) Historical and Scientific Survey of York and District. British Association for the Advancement of Science. York and London

Reptilia (Reptiles)

Photo: Richard K. Parsons

See amphibia/amphibians.

Chapter 5
Bacteria (and Related Organisms)

Bacteria As far as it can be ascertained there are no "comprehensive" or relatively comprehensive accounts of the bacteria and related organisms of any city in Europe. It is a grave omission.

© Springer International Publishing Switzerland 2016
J.G. Kelcey, *Provisional Bibliography of Atlases, Floras and Faunas of European Cities: 1600–2014*, DOI 10.1007/978-3-319-31120-3_5

Chapter 6
Fungi and Lichenised Fungi

Fungi

<u>Belgium</u>

Photo: John G. Kelcey

Bruxelles (Brussels)

Vanholen B, de Kesel A (2000) Inventarisatie en monitoring van de mycoflora en de lichen van het Brussels Hoofstededlijk Gewest. Jaarraport Werkgroep mycology (Inventory and monitoring of the mycoflora and lichens of the Brussels Capital Region). Nationale Plantentuin van Belgie, Meise

Vanholen B, de Kesel A, Fraiture A (2001) Inventarisatie en monitoring van de mycoflora en de lichenen van het Brussels Hoofdstedelijk Gewest. Eindrapport mycologische werkgroep. Intern rapport Nationale Plantentuin van Belgie/Brussels Instituut voor Milieubeheeer (Inventory and monitor-

© Springer International Publishing Switzerland 2016

J.G. Kelcey, *Provisional Bibliography of Atlases, Floras and Faunas of European Cities: 1600–2014*, DOI 10.1007/978-3-319-31120-3_6

ing of mycoflora and lichens of the Brussels-Capital Region. Final mycological group. Intern National Botanic Garden of Belgium/Brussels Institute for Milieubeheeer report), 56 p + CD-ROM

England

Halifax

Bolton J (1786-91) History of Fungusses growing about Halifax vols. 1–3

Poland

Lodz

Lawrynowicz M (1982) Macro-fungal Flora of Lodz. In: Bornkamm R, Lee JA, Seaward MRD (eds) Urban Ecology 2nd European Ecological Symposium. Blackwell, Oxford, pp. 41–47

Lichenised Fungi

Austria

Wien (*Vienna*)

Photo: John G. Kelcey

Krisai-Greilhuber I (1992) Die Makromyceten im Raum von Wien (The macromycetes in the space of Vienna). Okologie under Floristik. IHW, Eching

Turk R (2006) Die Flechtenflora in Wien-Veranderungen im Zeitraum 1853 bis 2004. (The Lichen Flora in Vienna – changes in the time-period 1853 to 2004). Verhandlungen de Zoologisch-Botanischen Gesellschaft in Osterreich, p. 143

Belgium

Bruxelles (*Brussels*)

Vanholen B, de Kesel A (2000) Inventarisatie en monitoring van de mycoflora en de lichen van het Brussels Hoofstededlijk Gewest. Jaarraport Werkgroep mycology (Inventory and monitoring of the mycoflora and lichens of the Brussels Capital Region). Nationale Plantentuin van Belgie, Meise

Vanholen B, de Kesel A, Fraiture A (2001) Inventarisatie en monitoring van de mycoflora en de lichenen van het Brussels Hoofdstedelijk Gewest. Eindrapport mycologische werkgroep. Intern rapport Nationale Plantentuin van Belgie/Brussels Instituut voor Milieubeheeer (Inventory and monitoring of mycoflora and lichens of the Brussels-Capital Region. Final mycological group. Intern National Botanic Garden of Belgium/Brussels Institute for Milieubeheeer report), 56 p + CD ROM

England

Birmingham

Lindsay DC (1981) Lichens of the Birmingham Region. Proceedings of the Birmingham Natural History Society 24: 125–152

Lindsay DC (1981) Birmingham and Warwickshire Lichens: new records. Proceedings of the Birmingham Natural History Society 24: 194–198

London

Laundon JR (1967) A Study of the Lichen Flora of London. Lichenologist 3: 277–327

Laundon JR (1970) London's Lichens. London Naturalist 49: 20–69

Italy

Roma (*Rome*)

Nimis PL (1989) Urban Lichen studies in Italy III. The City of Rome. Braun-Blanquetia 3: 279–286

Russian Federation

St. Petersburg

Malysheva N (2003) Lichens of St. Petersburg. St.Petersburg State University. Trudi Sankt-Peterbrgskogo Obschestva ispitatelei prirody 79(3) (in Russian). 97 pages

Малышева, Н. (2003) Лишайники Санкт-Петербурга. Издательство С.-Петербургского университета (Труды Санкт-Петербургского общества естествоиспытателей, т. 79. Серия 3). 97 с.

Chapter 7
Plants (Non-vascular)

Algae

Photo: John G. Kelcey

There appears to be no comprehensive or relatively comprehensive accounts of the algal flora of any European city.

© Springer International Publishing Switzerland 2016

J.G. Kelcey, *Provisional Bibliography of Atlases, Floras and Faunas of European Cities: 1600 –2014*, DOI 10.1007/978-3-319-31120-3_7

Bryophyta (Bryophytes)

Austria

Photo: John G. Kelcey

Wien (*Vienna*)

Zechmeister HG, Humer K, Hohenwallner D (1998) Historische Moosflora von Wien. Teil 1: Leber-und Hornmoose (Hepaticae, Anthocerotae) (Liverworts and Hornworts). Verh Zool Bot Ges Wien 135: 343–351

Zechmeister HG, Humer K, Hohenwallner D (1998) Historische Moosflora von Wien. Teil 1 Leber-und Hornmoose (Hepaticae, Anthocerotae) (Past moss flora of Vienna. Part 1 Liverworts and Hornworts (Hepatica, Anthocerotae)). Verh Zool Bot Ges Wien 135: 343–351

Zechmeister HG, Hohenwallner D, Humer K (1998) Historische Moosflora von Wien. Teil 2. Laubmoose (Musci) (Historic Moss Flora of Vienna. Part 2. Laubmoose (Musci)) Verh Zool Bot Ges Wien 135: 355–379

Zechmeister HG, Hohenwallner D, Humer-Hochwimmer K (2001) Die Erforschung de Moosflora von Wien. (Researching the Moss flora of Vienna). Berichte der Rheinhold-Tuxen-Gesellschaft (RTG) 13: 291–295

Belgium

Bruxelles (*Brussels*)

Not known whether published in French

Vanderpoorten A (1997) A Bryological Survey of the Brussels Capital Region (Belgium). Scripta Botanica Belgica 14

<u>England</u>

Ashton-under-Lyne (Greater Manchester)

Holt CA (1888) Hepatics. In: Whitehead J (1888) Mosses in the District of Ashton-under-Lyne

Whitehead J (1888) Mosses in the District of Ashton-under-Lyne – also see Plants (vascular)

Liverpool

Marrat FP (1860) On the Musci and Hepaticae of Liverpool and vicinity. Proc Liverpool Literary and Philosophical Society, Appendix 1–7

Marrat FP (1861) Additions to the Musci and Hepaticae of Liverpool and vicinity. Proc Liverpool Literary and Philosophical Society 14: 14–16

Marrat FP (1862) Additions to the Musci and Hepaticae of Liverpool and vicinity. Proc Liverpool Literary and Philosophical Society 15: 158

London

Peterken FHG (1961) A hand list of the plants of the London Area. Bryophytes London Naturalist 40: 43–71

Southport (also see Liverpool)

Marrat FP (1864) The Mosses of Southport. Liverpool Naturalists' Scrapbook 1863-64

<u>Italy</u>

Gela (Sicilia)

Polizzi MD, Lo Giudice R, Maugeri G (2005) Flora vascolare e briofitica della citta di Gela (Sicilia meridionale) (Bryophyte and vascular flora of the city of Gela (Southern Sicily)). Webbia 60

<u>Slovakia</u>

Bratislava

Janovicova K, Kubinska A, Javorcikova D (2003) Pecenovky (Hepatophyta), rozteky (Anthoceratophyta) a machy (Bryophyta) na uzemi Bratislavy (Slovensko) (Liverworts (Hepatophyta), Hornworts (Anthoceratophyta) and Mosses (Bryophyta) in the area of Bratislava (Slovakia)). Bot ustav Bratislava

Chapter 8
Plants (Vascular including Pteridophytes)

Photos: John G. Kelcey

© Springer International Publishing Switzerland 2016

J.G. Kelcey, *Provisional Bibliography of Atlases, Floras and Faunas of European Cities: 1600–2014*, DOI 10.1007/978-3-319-31120-3_8

Photo: John G. Kelcey

<u>Austria</u>

Bad Aussee

Rechinger L (1966) Die Flora von Bad Aussee. (The Flora of Bad Aussee) Graz

Eisenerz

Wagner R, Mecenovic K (1973) Flora von Eisenerz und Umbebung (Flora of Eisenerz and surroundings) Mitt. Abt.Bot. Landesmus. Joanneum Graz 2/3

Kloch

Maurer W, Mecenovic K (1970) Die Flora von Kloch und St. Anna am Aigen (Flora of Kloch and St. Anna am Aigen) Mitt Abt Zool Bot Landesmus. Joanneum Graz 37: 127–172

Krumegg

Maurer W (1974) Die Flora von Krumegg und St. Marien bei Graz (Steiermark, Osterreich) (The Flora of Krumegg and St. Marien near Graz) Mitt. Naturewiss. Ver. Steierm 104

Pischelsdorf

Maurer W, Mecenovic K, Pittoni-Dannenfeld (1975) Die flora von Pischelsdorf and Stubenberg (Steiermark, Osterreich) (The Flora of Pischelsdorf and Stubenberg (Styria, Austria)) Mitt. Abt.Bot. Landesmus Joanneum Graz 6

Salzburg

Wittmann HA, Ziebenbranner A, Pilsl P, Heiselmayer P (1987) Verreitungsatlas dar Salzburger GefaBpflanzezen (Chorological Atlas of the vascular plants of Salzburg) Sauteria 2: 1–403

Wien (*Vienna*)

Janchen E (1977) Flora von Wien. Niederosterreich und Nordburgenland. (Flora of Vienna)Verein fur Landeskunde von Niederosterreich und Wien

Adler W, Mrkvicka AC (2003) Die Flora Wiens gestern und heute: Die wildwachsenden Farn- und Blütenpflanzen in der Stadt Wien von der Mitte des 19. Jahrhunderts bis zur Jahrtausendwende (The wild ferns and flowering plants in the city of Vienna from the mid-19th century to the turn of the millennium) Naturhistorisches Museum, Wien

Belgium

Bruxelles (*Brussels*)

Holland F (1976) Contribution a l'etude de l'ecologie urbaine. Apercu de la Flore et de la Vegetation Bruxelloise. (Contribution to the study of urban ecology. Overview of the Flora and Vegetation of Brussels). Mem Fac Sci Bot Bruxelles

IBGE (1999) Atlas de la flore de la Region de Bruxelles-Capitale (Atlas of the flora of the Brussels-Capital Region). Institut Bruxellois pour la Gestion de l'Environnement Brussels. (Brussels Institute for Management of the Environment of Brussels)

Van Landuyt W, Hoste I, Vanhecke L, Van Den Bremt P, Vercruysse W, de Beer D (2006) Atlas van de flora van Vlaanderen en het Brussels Gewest (Atlas of the flora of Flanders and Brussels). Flo Wer/Instituut voor Natuur- en Bosonderzoek/Nationale Plantentuin van België, Brussel. ISBN 90-726-1968-4

Bulgaria

Sofia

Dimitrov DS (2004) The Vascular Flora of the City Sofia. In: Penev L, Niemelä J, Kotze D J, Chipev N (eds) Ecology of the City of Sofia, Species and Communities in an Urban Environment, Pensoft, Sofia-Moscow: 185–208

Croatia

Rovinj & Krk

Schulte W (1989) Zur Flora und Vegetation der Stadte Rovinj und Krk (Jugoslavien) (The flora and vegetation of Rovinj and Krk) Tuexenia 9: 199–223

Czech Republic

Horazdovice

Mandak B (1996) Přispěvek k floře Horaždovic. (Contribution to the flora of Horažďovice). Zpr Čes Bot Spolec Praha: 30 (1995) 127–134

Plzeň (Pilsen)

Nesvadbova J, Sofron J (1997) Flora a vegetace města Plzně (Flora and Vegetation of Pilsen). Zapadočeské Museum, Plzeň

Praha (Prague)

Špryňar P, Münzbergová Z (1998) Prodromus pražské květeny (Treatise on the Prague Flora) Muzeum a Současnost, 12: 129–222

Eire

Dublin

Jackson PW, Skeffington MS (1984) Flora of Inner Dublin. Royal Dublin Society Dublin

Doogue D, Nash D, Parnell J, Reynolds S, Wyse-Jackson PS (eds) (1998) Flora of County Dublin. Dublin Naturalists Field Club Dublin

England

Andover

Clarke CB (1866) A List of the Flowering Plants in the immediate vicinity of Andover

Ashton-under-Lyne (now in Greater Manchester)

Holt GA (1888) Flora of Ashton-under-Lyne and District, including bryophytes. (Note: the author is also given as Linnean Botanical Society of Ashton-under-Lyne)

Whitehead J (1888) The Flora of Ashton-under-Lyne

Collier F (1914) Additions to the Flora of Ashton-under-Lyne. L Nat (1914)

Basildon

Cole RL, Peck HJ, Reid G (1974) The Natural History of Basildon. Basildon Council

Bath

Gibbes H (date not known but before 1831) Flora Bathon

Babington CC (1834) Flora Bathoniensis

Babington CC (1839) Flora of Bathon. Supplement

Biddulph

Painter WH (1892) The Botany of Biddulph. Midland Naturalist 15: 131–9, 162–4 and 183–9

Birmingham

Trueman J, Poulton, M and Reade P (2013) Flora of Birmingham and the Black Country. Pices Publications, Newbury

Bishop's Stortford

Anon (1952) List of Vascular Plants of Bishops Stortford and District. Bishops Stortford and District Natural History Society

Bolton (now in Greater Manchester)

Greenless T, Holden TK (1920) The Flora of Bolton. Lancashire and Cheshire Naturalist 1920–1

Bournemouth

Linton EF (1900) Flora of Bournemouth. Privately published

Anon (1914) The Natural History of Bournemouth and District. Bournemouth Natural Science Society

Linton EF (1919) Appendix 1 to the Flora of Bournemouth

Linton EF (1925) Appendix 2 to the Flora of Bournemouth

Chapman Ray (ed) (2010) The Natural History of Bournemouth and the Surrounding Area

Bradford

Lees FA (1874) On the Flora of Leeds and Bradford District. J Bot III

Bristol

Swete EH (1854) Flora Bristoliensis

White JW (1912) The Flora of Bristol. John Wright and Sons Ltd, Bristol. Reprinted 1972 by Chatford House Press, Bristol

Willis AJ (1988) Bristol Botany in 1988. Proceedings of the Bristol Naturalists' Society 48: 21–31

Myles S L (ed) (2000) The Flora of the Bristol Region. Pisces Publications Newbury

Burton-on-Trent

Anon (1899-1903) The Flora of Burton-on-Trent and neighbourhood. Transactions of the Burton-on-Trent Natural History and Archaeological Society 3: (177–190, 269 and 282); 4: (75–85 and 117–148); 5: 37

Cambridge

Martyn T (1763) Planta Cantabrigiensis

Christchurch

Woodhead F (1994) Flora of the Christchurch area

Cirencester

Greenwood WJ (1914) The Flora of Cirencester and its Neighbourhood. Cirencester Naturalist and Archaeologist Club

Derby

Futter K, Raynes P (1989) The Flora of Derby. Futter and Raynes, Derby

Dewsbury

Lee PF (1887) Flora of Dewsbury and Neighbourhood. Transactions of the York Naturalists' Union for 1885, pub. 1887

Doncaster

White Dr (1804) List of Plants of York and Doncaster. In: Miller E (ed) History and Antiquities of Doncaster, Doncaster

Halifax

Bolton J (attributed) (1775) Catalogue of Plants Growing in the Parish of Halifax. In: Watson J (ed) History and Antiquities of the Parish of Halifax

Crump WB, Crossland C (1904) Flora of the Parish of Halifax. Note: this reference is also quoted as "Crossland C (1904) Flora of Halifax."

Murgatroyd F (1988) Flowering Plants of the Halifax Parish, Halifax Scientific Society

Murgatroyd F (1989) Flowering Plants of the Halifax Parish. Halifax Scientific Society

Hull (the common name; the correct name is Kingston upon Hull)

Middleton R (1998) The Plants of Hull: an Electronic Atlas. Naturalist 123: 24–6

Middleton R (2001) The Plants of Hull: a Millennium Atlas. Hull Natural History Newsletter Supplement 2

King's Lynn

Payne RM (1995) Flora of King's Lynn. Transactions of the Norfolk and Norwich Naturalists' Society 30: 317–342

Leeds

Lees FA (1874) On the Flora of Leeds and Bradford District. J Bot III

Nelson GA (ed) (1963) A Flora of Leeds and District. Proceedings of the Leeds Philosophical and Literary Society (Scientific Section) 9: 113–178

Leek

Pettty L (1893) Plants of Leek and Neighbourhood.

Liverpool

Hall TB (1839) Flora of Liverpool

Dickinson J (1851) Liverpool Flora. (2nd edn)

Dickinson J (1855) Supplement to the Floras of Liverpool. Proceedings, Liverpool Literary and Philosophical Society, 9

Fisher HS (1861) Liverpool Plants. Proceedings of the Liverpool Literary and Philosophical Society, 15

Dickinson J (1872) Liverpool Flora. (3rd edn). Liverpool Naturalists' Field Club

Fisher HS (ed) (1872) Flora of Liverpool

Brown R (1873) Appendix to the Flora of Liverpool. Liverpool Naturalists' Field Club

Brown R (1875) Second Appendix to the Flora of Liverpool. Liverpool Naturalists' Field Club

Brown R (1887) Third Appendix to the Flora of Liverpool. Liverpool Naturalists' Field Club

Brown R (1896) Botany of the Liverpool District. British Association Handbook to Liverpool

Green CT (ed) (1902) Liverpool Flora. (4th edn)

Green CT (ed) (1933) The Flora of the Liverpool District, Ed. 2. T. Buncle, Arbroath

Savidge JP (1958) Flora of the Liverpool Area 1932-1957. Liverpool Naturalists' Field Club Proceedings 1957: 17–26

London

Gerard J (1597) The Herball, or general historie of plantes, gathered by John Gerard of London, Master in Chirurgerie. John Norton, London. (Note: this reference is also given as Gerard J (1957) The Herball or General Historie of Plantes. Bollifont, London)

Curtis W (1775-1798) Flora Londinensis (also given as 1771)

Hooker WJ (1816-1828) Flora Londinensis

Cooper D (1836) Flora Metropolitana or Botanical Rambles within thirty miles of London. S. Highley, London

Irvine A (1838) The London Flora. Smith, Elder and Co. London

de Crespigny E (1877) A New London Flora or handbook of the botanical localities of the Metropolitan Districts. Hardwicke and Bogue, London

Kent DH, Lousley JE (1951-57) A Hand List of the Plants of the London Area. Supplement to the London Naturalist: 30-36

Bangerter EB (1961) The Botany of the London Area. London Naturalist 40: 6–16

Burton RM (1983) Flora of the London Area. London Natural History Society, London

Edgington JA (2003) Ferns of the Metroplis—a status report. London Naturalist 82: 59–73

Luton

Dony JG, Dony CM (1991) The Wildflowers of Luton. J.G. Dony

Manchester (Greater Manchester)

Wood JB (1840) Flora Mancuniensis

Buxton R (1849) Botanical Guide to the Flowering Plants within sixteen miles of Manchester

Grindon LH (1859) The Manchester Flora

Grindon LH (c.1860) Manchester Walks and Wildflowers Combert (1872) Manchester Plants J Bot 10: 376

Middleton (now in Greater Manchester)

Shaw C (1989) The Plants of Middleton

Shaw C, Hill P, Bescoby E (1989) Some Natural History of Middleton

Mildenhall

Bunbury CJF (1899) Flowering Plants and Ferns of Mildenhall

Newbury

Anon (1839) Newbury and its environs—A Catalogue of Plants found in the neighbourhood. Hall and Marsh. (Note: this reference is also given as the "Flora of the Newbury Area")

Nottingham

Shepherd P (1998) The Plants of Nottingham—A City Flora. Wildtrack Publishing, Sheffield

Oldham

Langridge B (1996) The Wildflowers of Oldham. Oldham Education and Leisure, Oldham

Oswestry

Diamond TP (1891) A Flora of Oswestry and District. Woodall Minshall & Co., Oswestry

Oundle

Rees JS (1970) A Flora of Oundle. Oundle School Natural History Society

Oxford

Sibthorp J (1794) Flora Oxoniensis. Oxonii; Typis Academicis

Plymouth

Briggs TRA (1880) Flora of Plymouth. Pub John Van Voorst, London

Stevens RA (1990) A Provisional Flora and Habitat Atlas of Plymouth. Nature Conservancy Council

Preston

Ashfield CJ (1858-65) Flora of Preston and Neighbourhood. Trans Historical Society Lancashire and Cheshire

Preston Scientific Society (1903) Flora of Preston and Neighbourhood

Reading

Attwood JS, Gilmour JSL (1949) TB Flower's List of Reading Plants. Watsonia 1: 139

Reigate

Luxford (1838) Flora of Reigate
Brewer (undated; pre 1877) New Flora of Reigate

Rochdale (now in Greater Manchester)

Shaw C, Pentelow W (1972) Flowers of the Rochdale Area

Royston

Kingston A (1883) Flowering Plants at Royston in 1882. Trns Herts Nat Hist Soc 2

Rugby

Allen DE (1948) The Rugby Flora 1946-47. Report of the Rugby School Natural History Society 81: 18–30
Allen DE (1957) The Flora of the Rugby District. Report of the Rugby School Natural History Society

Saddleworth (now in Greater Manchester)

Kidd LN, Edwards WF (1977) A Flora of Saddleworth

Sheffield

Salt J (1800) Flora Sheffieldiensis
Snelgrove E (1894) Flora of the Sheffield District. Report of the Sheffield Naturalists' Club
Snelgrove E (1895) Second List of the Flora of the Sheffield District. Report of the Sheffield Naturalists' Club
Snelgrove E (1910) Botany of the District * Proc Vol 1 SNC Report 1910
(* = Sheffield)
Shaw M (ed) (1988) A Flora of the Sheffield Area (Two hundred years of plant records) Sorby Record Special Series 8. Sorby Natural History Society, Sheffield

Todmorden

Standsfield A (1911) Flora of Todmorden

Torquay

Stewart R (1860) Handbook of the Torquay Flora

Tunbridge Wells

Forster TF (1816) Flora Tonbridgensis; or a Catalogue of Plants growing wild in the Neighbourhood of Tonbridge Wells, arranged according to the Linnaean System, London

Forster TIM (1842) Flora Tonbridgensis; or a Catalogue of Plants growing wild in the Neighbourhood of Tonbridge Wells, arranged according to the Linnaean System (2nd edn), London

Jenner E (1845) A Flora of Tunbridge Wells, being a list of indigenous plants within a fifteen mile radius. J Colbrook, Tunbridge Wells

Wakefield

Gissing TW (1862) The Ferns and Fern allies of Wakefield and its Neighbourhood

Gissing TW (1867) Flora of Wakefield. Reprinted from "The Naturalist 1865-66 Huddersfield"

Warrington

Robinson JF (1859) British Ferns about Warrington. Phytologist (New Series) 5: 99

Watford

Cottam A (1875) Notes on the Flora of the Watford District. Transactions of the Watford Natural History Society: 1

Weston-super-Mare

St. Brody GAO (1850) Flora of Weston-super-Mare

Widnes

Anon (1912) Flora of Widnes. Widnes and District Field Club

York

White Dr. (1804) List of Plants of York and Doncaster. In: Miller E (ed) History and Antiquities of Doncaster. Doncaster

Christy RM (1882-3) A List of Ferns in the Neighbourhood of York. Naturalist 8: 177–184

Ibbotson H (1884) The Ferns of York. William Sessions, York

Wilkinson HJ (1906) Phanerogamic Flora and Vascular Cryptogams. In Auden GA (ed) Historical and Scientific Survey of York and District. British Association for the Advancement of Science, York and London

Merryweather J (1995) City Ferns. Yorkshire Philosophical Society Annual Report 1995

Finland

Helsinki

Kurtto A, Helynranta L (1998) Helsingin kasvit, Kukkivilta kivilta metsan syliin. (Flora of Helsinki—from lowering stones to forest floor). City of Helsinki Environment Centre and Helsinki University Press

France

Paris

Vaillant S (1727) *Botanicon Parisiense*. Verbeek and Lakeman, Leiden and Amsterdam

Germany

Altenburg

Strumpf K (1969) Flora von Altenburg unter besonderer Berucksichtigung der Entwicklung des Artenbestandes von 1768-1968 (Flora of Altenburg with special reference to the development of the species inventory of 1768-1968) Abh u Ber Naturdl Mus "Mauritianum" 6: 93–161.

Strumpf K (1992) Flora von Altenburg (Flora of Altenburg) Mauritiana 13: 339–523

Altdorf and Gießen

Jungerman L (1615) Katalogus plantarum quae area Altorfium norcum et vicinus quibusdam locis. Conrad. Agricola, Altorfi

Augsburg

Alten JW (1822) Augsburgische Blumenlese (Flora of Augsburg). J. Wolffische Buchhandlung, Augsburg

Hiemeyer F (ed) (1978) Flora von Augsburg (Flora of Augsburg) Ber. Naturwiss ver Schwaben, Special Edition

Stangl J (ed) (1985) Pilzflora von Augsburg und Umgebun (Fungal flora of Augsburg and vicinity) Pilzverein Augsburg

Berlin

Elssholtz JS (1657) Hortus Berolinensis. Berlin

Hecker J (1742) Specimen florae Berolinensis. Berlin (Reines Kartenwerk) (Pure cartography)

Hecker JJ (1757) Flora Berolinensis. Das ist, Abdruck der Kräuter und Blumen nach der besten Abzeichnung der Natur, zur Beförderung der erkenntniss des Pflanzen-Reiches veranstaltet von der Real-Schule in Berlin. Centuria 1 – Berlin (Real-Schulbuchhandlung) (The flora of Berlin. That is, an imprint of herbage and flowers following the best copy of nature, for the improvement of knowledge of the plant kingdom edited by the Real-School in Berlin)

Willdenow CL (1787) Florae Berolinensis prodromus.(Treatise on the Flora of Berlin) Vieweg, Berolini. Note: reprinted 1987

Von Schlechtendal DFL (1823 and 1824) Flora Berolinensis. 2 vols. Dummler, Berlin

Dietrich A (1824) Flora der Gegend um Berlin (Flora of the area around Berlin), Berlin

Lackowitz W (1868) Flora von Berlin und der Provinz Brandenburg (Flora of Berlin and Province of Brandenburg). Friedberg & Mode, Berlin

Lackowitz W (1922) Flora von Berlin und der Provinz Brandenburg (Flora of Berlin and Province of Brandenburg). Friedberg & Mode, Berlin

Bocker R, Auhagen A, Brockmann H, Kowarik, I, Scholz H, Sukopp H, Zimmermann F (1991) Liste der wildwaxhsenden Farn- und Blutenpflanzen von Berlin (west) (List of the Wild-growing Ferns and Flowering Plants of (west) Berlin). In: Auhagen A, Platen R, Sukopp H (eds) Rote Listen der gefahrdeten Pflanzen und Tiere in Berlin 1990 (Red Lists of Endangered Plants and Animals in Berlin 1990). Landschaftsentwicklung und Umweltforschung 6: 57–58

Prasse R, Ristow M, Klemm G, Machatzi B, Raus TH, Scholz H, Stohr G, Sukopp H, Zimmermann FR (2001) Liste der wildwachsenden Gefäßpflanzen des Landes Berlin mit Roter Liste - Senatsverwaltung für Stadtentwicklung/Der Landesbeauftragter für Naturschutz und Landschaftspflege (List of wild vascular plants of Berlin with the Red List—Senate Department for Urban Development/The State Commissioner for Conservation and Landscape Care)

Seitz B, Ristow M, Prasse R, Machatzi B, Klemm G, Bocker R, Sukkop H (2012) Der Berliner Florenatlas (Atlas of the Flora of Berlin). Verhandlugen des Botanischen Vereins von Berlin und Brandenburg

Braunschweig

Brandes D (1984) Die Flora von Braunschweig um 1650 im Spiegel de "*Index planarum*" von Johann Chemnitius (The Flora of Braunschweig in 1650 in the mirror of "Index planarum" by Johann Chemnitius). Braunschw Naturk Schr 2: 1–18

Brandes D (1987) Verzeichnis der im Stadtgebiet von Braunschweig wildwachsenden und verwilderten Gefäßpflanzen (Directory of the city of Braunschweig wild and feral vascular plants) Braunschweig. Universitäts-Bibliothek der Technischen Universität Braunschweig

Chemnitz

Grundmann H (1992) Die wildwachsnden und verwilderten GefaBplanzen der Stadt Chmenitz und seiner unmittelbaren Umgebung (The wild and escaped ferns and flowering plants of the city of Chemnitz and its immediate surroundings). Veroffentlichengen des Museums fur Naturkunde Chemnitz 15

Danzig

See Gdansk

Darmstadt

Klaus-Dieter J (1992) Flora des Stadgebites von Darmstadt. Erebnisse einer Rasterkartierung. (Flora of the city area of Darmstadt. Results of a grid mapping). Sonderband der Berichte des Naturwissenschaftlichen Veeins, Darmstadt

Dessau

Voigt O (1982) Flora von Dessau und Umgebung (The Flora of Dessau and surroudings) 2 Teil. Naturwissenschaftliche Beitrage des Museums Dessau. Sonderheft. pp. 1–96

Duisburg

Dull R, Kutzelenigg H (1980) Punktkartenflora von Duisburg und Umgebung (Point cards Flora of Duisburg and the surrounding area). Westdeutscher Verl. Opladen

Essen

Reidl K, Dettmar J (1993) Flora und vegetation der Stadte des Ruhrgebietes, insbesondere der Stadt Essen und der Inustrieflachen (Flora and vegetation of the Ruhr area cities, especially the city of Essen and the industrial areas). Ber. Dt. Landeskunde 67: 299–326.

Euskirchen

Zimmermann-Pawlowsky A (1985) Flora und vegetation von Euskirchen und ihre Veranderung in den letzten 70 Jahren (Flora and vegetation of Euskirchen and their alteration in the last 70 years). Decheniana 138: 17–37

Frankfurt am Main

önsel D, Ottich I, Malten A, Zizka G (2008) An updated list of the vascular plants of Frankfurt am Main (Pteridophyta & Spermatophyta). *Senckenberg Bio* 88, 111–121

Göttingen

Garve E (1985) Artenliste und Anmerkungen zur rezenten Gefasspflanzenflora der Stadt Gottingen. (Species list and remarks on recent flora of ferns and flowering plants of the city of Göttingen). Mit Fauna u Flora Sudniedersachsens 7: 163–179

Hamburg

Poppendieck HH, Bertram H, Brandt I, Engelschall B, von Prondzinski J (2010) Der Hamburger Pflanzenatlas von a bis z (The Hamburg Plant Atlas from A to Z) Dolling und Galitz, Munchen & Hamburg

Leipzig

Gutte P (1989) Die wildwachsenden und verwilderten Gefäßpflanzen der Stadt Leipzig (The wild and feral flowering plants of the city of Leipzig) Veroff. Naturkundemus 7: 1–95

Munchen (*Munich*)

Kreuzpointer JB (1876) Notizen Zur Flora Munchens. (Notes on the Flora of Munich). Flora Jena 59: 77–80

Osnabruck

Overdieck D, Scheitenberger A (1989) Veränderungen des Arteninventars der Vegetation in einer mitteleuropäischen Stadt (Osnabrück) (Changes in the plant spectrum of the vegetation of a Central European city (Osnabruck)). Verh. Ges. Ökologie 1988: 265–271

Regensburg

Mergenthaler O (1982) Verbreitungsatlas zur Flora von Regensburg. (Distribution atlas for the flora of Regensburg) Hoppea Denkschr Regensb Bot Ges 40: V–XII, 1–297

Saarlouis

Maas S (1983) Die Flora von Saarlouis (The Flora of Saarlouis). Abh Delattinia 13: 1–108

Stuttgart

Kirchner A (1888) Flora von Stuttgart und Umgebung (Ludwigsburg, Waiblingen, Esslingen, Nurtingen, Leonberg, ein Teil des Schobuches) mit besondered Berucksichtigung der pflanzenbiologischen Verhaltinisse (Flora of Stuttgart and surroundings (Ludwigsburg, Waiblingen, Esslingen, Nurtingen, Leonberg) with special hints of the plant biological circumstances).Eugen Umer, Stuttgart

Wolfsburg

Griese D (1999) Flora und Vegetation einer neuen Stadt am Beispiel von Wolfsburg (Flora and vegetation of a new city on the example of Wolfsburg). Braunschw Geobot Arb 7: 1–235

Wuppertal

Stieglitz W (1987) Flora von Wuppertal (Flora of Wuppertal) Jahresber. Naturwiss. Vereins Wuppertal, Beiheft 1: 1–227

Greece

Alexandroupolis

Chronopoulos G, Christodoulakis D (2006) Contribution to the Urban ecology of Greece: The flora of Alexandroupolis (NE Greece) and its vicinity. *Fresen. Environ. Bull.* 15, 1455–1466

Mesolongi

Tsiotsiou V, Christodoulakis D (2004) Contribution to the urban ecology of Greece: The flora of Mesolongi city. *Fl. Medit.* 14, 129–151

Patras

Chronopoulos G, Christodoulakis D (1996) Contribution to the urban ecology of Greece: The flora of the city of Patras and the surrounding area. Bot Helv 106: 159–176

Thessaloniki

Krigas N, Kokkini S (2004) A survey of the alien vascular flora of the urban and suburban area of Thessaloniki, N Greece. *Willdenowia* 34, 81–99

Krigas N, Kokkini S (2005) The indigenous vascular flora of the urban and suburban area of Thessaloniki (N. Greece). *Bot. Chron.* 18, 29–84

Italy

Ancona

Hruska K (1988) Flora e vegetazione della citta di Ancona (Flora and Vegetation of Ancona). Giorn Bot Ital 122 (1–2), suppl. 1: 67

Aosta

Filipillo S, Sartori F, Terzo V, Gardini PS (1986) Carta della vegetazione dei dintorni della citta di Aoasta (Vegetation map of the vicinity of the town of Aosta) Atti Ist Bot Lab Critt Univ Pavia serie 7, 4 (1985) 5–25

Camerino

Hruska K (1996) Carta della vegetazione della citta di Camerino (MC) (Vegetation map of the town of Camerino) Giorn Bot Ital 130 (1)

Chieti

Leporatti ML, Pavesi A, Massari G (2001) Contributo alla conoscenza della flora vascolare urbana di Chieti (Abruzzo) (Contribution to the knowledge of the vascular flora of urban Chieti). Webbia 56(2): 343–378

Cogoleto

Barberis G, Bertolotto S, Peccenini S (1993-1994) Flora urbana di Cogoleto (GE) (Urban Flora of Cogoleto). Allionia 32: 287–301

Firenze (Florence)

Arrigoni PV, Rizzotto M (1993-1994) Caratteri della flora e della vegetazione urbana di Firziene (Character of the flora and vegetation of urban Florence). Allionia 32: 231–243

Gela (Sicilia)

Polizzi MD, Lo Giudice R, Maugeri G (2005) Flora vascolare e briofitica della citta di Gela (Sicilia meridionale) (Bryophyte and vascular flora of the city of Gela (Southern Sicily)). Webbia 60

Genova (*Genoa*)

Barberis G, Cevasco R, di Turi A (1994). Alcune considerazioni sulla distribuzione delle specie vegetali nella citta di Genova (Some considerations on the distribution of plant species in the town of Genoa) Gion Bot Ital 128: 203

Milano (*Milan*)

Banfi E, Galasso G (1998) La flora spontanea della cita di Milano alle soglie del terzo millennio e i suoi cambiamenti a partire dal 1700 (The natural flora of the city of Milan at the threshold of the third millennium and its changes since 1700). Mem Soc Ital. Sci Nat Mus Civico Storia Nat. Milan 27 (3)

Napoli (*Naples*)

Tenore M (1811-38) Flora Napolitana (Flora of Naples)

De Natale A, La Valva V (2000) La Flora di Napoli: i quartieri della citta (Flora of Naples: neighbourhoods of the city) Webbia 54: 271–375

Ortona

Pirone G, Corbetta F, Raffaelli B (2002) La flora urbica della citta di Ortona (Chieti, Abruzzo) (Urban flora of Ortona) Quad. Bot. Amb. Appl. 13: 27–40

Roma (*Rome*)

Celesti-Grapow L (1985) Atlante della Flora di Roma (Atlas of the Flora of Rome). Sapienza University, Comune di Roma, Argos ed Roma

Celesti-Grapow L (1995) Atlante della Flora di Roma (Atlas of the Flora of Rome). Comune di Roma, Argos Edizione, Roma

Scordia (*Sicilia*)

Lo Giudice R, Palmieri R, Gueli L, Maugeri G (2005) Flora urbana della citta di Scordia (Sicilia orientale) (Urban Flora of the city of Scordia, Eastern Sicily). Webbia 60

Trento

Pedrotti F (1992) Observations preliminaries sur la flore et la vegetation de ville de Trente (Italie du Nord) (Preliminaries observations on the flora and vegetation of Trento (northern Italy)). Braun-Blanquetia 3 (1989) 121–126

Vercelli

Pomini L (1986) Flora urbica della citta di Vercelli ed adiacenze (Urban Flora of Vercelli and its vicinity) Instito Tecnuo Agraco, Vercelli

Northern Ireland

Belfast

Tate R (1863) Flora belfastiensis.

Beesley S, Wilde J (1998) Urban Flora of Belfast. Institute of Irish Studies, The Queen's University of Belfast, Belfast

Poland

Gdansk (formerly Danzig)

Klinsmann E (1836) Beiträge zur Flora von Danzig nebst Umgegend. T. 1, 2, 3 (Contributions to the Flora of Danzig, and the adjacent area). Neumanns Stadtdruckerei, Gleiwitz einigen andern Preussischen Gegenden. Preuss. Prov. Bl. 15: 467–477

Gliwice, Koźle, Opole

Jungck M (1889-1890-1891) Flora von Gleiwitz und Umgegend. 1, 2, 3. (Flora of Gliwice Town and the adjacent area). Neumanns Stadtdruckerei, Gleiwitz einigen andern Preussischen Gegenden

Glogowek

Szotkowski P (1987) Flora miasta Głogówka na Śląsku Opolskim (Flora of Głogówek in Opole Silesia). OTPN Opole

Jaworznie

Tokarska-Guzik B (1999) Atlas rozmieszczenia roślin naczyniowych w Jaworznie (Atlas of Vascular Plant Distribution in Jaworzno). Zeszyty Naukowe Uniwersytetu Jagiellońskiego, Prace Botaniczne 34: 1–292

Krakow (*Cracow*)

Berdau F (1859) Flora Cracoviensis Typis (Flora of Cracow). C R Universitatis Jagellonicae, Cracoviae

Legnica

Anioł-Kwiatkowska J (1974) Flora i zbiorowiska synantropijne Legnicy, Lubina i Polkowic (Flora and Synanthropic Vegetation of Legnica, Lubin and Polkowice). Acta Universitatis Wratislaviensis 229, Prace Botaniczne 19: 3–152

Łodz

Witosławski P (2006) Atlas rozmieszczenia roślin naczyniowych w Łodzi. (Atlas of the Distribution of Vascular Plants in Łódź). Wydawnictwo Uniwersytetu Łódzkiego, Łódź

Lubań Śląski

Weretelnik E (1979) Flora i zbiorowiska synantropijne Lubania Śląskiego (The flora and synanthropic communities of Lubań Śląski). Acta Universitatis Wratislaviensis 354, Prace Botaniczne 23: 15–80

Lublin

Anioł-Kwiatkowska J (1974) Flora i zbiorowiska synantropijne Legnicy, Lubina i Polkowic (Flora and Synanthropic Vegetation of Legnica, Lubin and Polkowice). Acta Universitatis Wratislaviensis 229, Prace Botaniczne 19: 3–152

Fijalkowski D (1994) Flora roslin naczyniowych Lubelszczyzny. 2 vols. (Flora of vascular plants of Lublin). Srodowisko Przrodnicze, Lubelszczyzny

Polkowice

Anioł-Kwiatkowska J (1974) Flora i zbiorowiska synantropijne Legnicy, Lubina i Polkowic (Flora and Synanthropic Vegetation of Legnica, Lubin and Polkowice). Acta Universitatis Wratislaviensis 229, Prace Botaniczne 19: 3–152

Poznan

Jackowiak Bogdan (1993) Atlas rozmieszczenia roślin naczyniowych w Poznaniu (Atlas of distribution of vascular plants in Poznań). Prace Zakładu Taksonomii Roślin Uniwersytetu A. Mickiewicza 2: 5–409

Rusinska A (2002) Wszedobylskie mchy I watrobowce (Mosses and liverworts ubiquitous) Kronika Miasta Poznania 3: 39–46

Warszawa (Warsaw)

Bernhardi M (1652) Catalogus plantarum tum exoticarum quam indigenarum quae anno MDCLI in hortis regiis Varsaviae et circa eandem in locis sylvaticis, pratensibus, arenosis et paludosis nascuntur collectarum. Gdańsk

Erndtel Ch H (1730) Warsavia physice illustrata sive de aere, aquis, locis et incolis Warsaviae eorundemque moribus et morbis tractatus, cui annexum est Viridiarum et Catalogus plantarum circa Warsaviam nascentium. Drezno

Hryniewiecki B (1954) Pierwsze flory okolic Warszawy (The First Warsaw Flora). Monographiae Botanicae 2: 1–76

Sudnik-Wójcikowska B (1987) Flora miasta Warszawy i jej przemiany w ciągu XIX i XX wieku. Część 1. Część 2. Dokumentacja. (Flora of Warsaw and the changes during the 19th and 20th century. Parts 1 and 2. (Documentation)) Wydawnictwa Uniwersytetu Warszawskiego, Warszawa

Sudnik-Wójcikowska B (ed.) (1998) Flora miasta Warszawy i jej przemiany w ciągu XIX i XX wieku. Część 3. Dokumentacja (1987-1997) (Flora of Warsaw and the changes during the 19th and 20th century. Part 3. Documentation (1987-1997)). Wydawnictwa Uniwersytetu Warszawskiego, Warszawa

Russian Federation

Izevsk

Reference in Cyrillic

Iliminskich NG, Baranova OG, Pusirev AN (1998) Flora of Izhevsk and vicinity. In: Nature of Izhevsk and its vicinity. Izhevsk. pp. 81–169

Lipetsk

Reference in Cyrillic

Nekrasova VL (1918) List of Plants of Lipetsk City of Tambovskoi Gubernii In Isvestiya Glavnogo Botanicheskogo Sada OF RSFSR Tom 18 (2): 17–26

Reference in Cyrillic

Nekrasova VL Aleksandrov LP (1921) Addition to the list of Lipetsk City of Tambovskoi. 20(1): 5–13

Moscow

Reference in Cyrillic

Shvetsov AN (1997) The synopsis of the flora in the area of Moscow. Bulletin of the Main Botanical Garden 174: 47–57

Флора Москвы/ Варлыгина Т.И., Головкин Б.Н., Киселева К.В., Майоров С.Р., Немченко Э.П., Новиков В.С., Швецов А.Н., Щербаков А.В.; под общей ред. проф. В.С. Новикова. /Департамент природопользования и охраны окружающей среды города Москвы. – М.: Голден-Би, 2007. – 512 с.,: ил.(Flora of Moscow. (2007). Varligina, T.I., Golovkin B.N., Kiseleva K.V., Maiorov S.P., Nemchenko E.P., Novokov V.S., Shvetsov A.N., Scherbakov A.N. Department of nature protection of Moscow. Moscow, Golden-Bee)

Petrozavodsk

Reference in Cyrillic

Antipina GS, Venzhik Yu.V, Toivonen IM (2001) Higher Vascular Flora of Petrozavodsk City

St. Petersburg

Соболевский, Г. Санктпетербургская флора или описание находящихся в Санктпетербургской губернии природных растений, с приложением некоторых иностранных, кои на открытом воздухе в здешнем странoположении удобно произрастают, и с показанием оных силы, действия и употребления, в пользу для сельских жителей и любителей травознания. В 2 ч. Ч. 1-2. СПб: Тип. при Губернском правлении., 1801-1802 (Sobolevsky G. (1801-1802) Flora of St. Petersburg. Part 1 and part 2. St. Petersburg-Publisher of the Provincial Government).

Reference in Cyrillic

Nekrasova N (1959) Flora of St. Petersburg and its nearest suburbs in the 20th century. Turt Bot J 44: 249–261

Saratov

Панин АВ, Березуцкий МА. Анализ флоры города Саратова. Ботанический журнал, том 92, номер 8, год 2007 (Panin A, Beresutsky M (2007) Analysis of Saratov Flora. Botanical Journal, Tom 92, Number 8)

Ulyanovsk

Reference in Cyrillic

Levina PV, Golitsin SV (1953) Flora of Ulyanovsk and vicinity and Opportunity for Use. In: Kraevedcheskie zapiski Ulyanovskogo oblastnogo kraevedcheskogo museya 1. C: 87–101.

Voronezh

Reference in Cyrillic

Grigorievskaya AG (2000) Flora of Voronezh City. Voronezh State University Publisher

Scotland

Aberdeen

Trail JWH (1923) Flora of the City Parish of Aberdeen. In: James William Helenus Trail, a Memorial Volume. Aberdeen University Press, Aberdeen. pp 57–331

Edinburgh

Smith PM, Dixon ROD, Cochrane MP (2002) Plant life of Edinburgh and the Lothians. University of Edinburgh Press, Edinburgh

Glasgow

Scott-Elliot GF, Laurie M, Murdoch JB (1901) Fauna, flora and geology of the Clyde area Glasgow. Local Committee for the meeting of the British Association

Dickson JH (1991) Wild Plants of Glasgow. Aberdeen University Press or Mercat Press, Edinburgh

Dickson JH, Macpherson P, Watson K (2000) The Changing Flora of Glasgow: Urban and rural plants through the centuries. Edinburgh University Press

Slovakia

Bratislava

Horvátovsky S (1774) Flora Tyrnaviensis indigena pars prima. Quam cum annexis thesibus etc. pro consequenda suprema medicinae laurea, et juribus ac privilegis eidem annexis die mensis Augusti Anni 1774 hora consueta in palatio universitatis publicae disquisitioni exponet Sigismundus Horvátovsky. Hungarus Bartphenensis e Comitatu Sárosiensi. I–IV + 1–48. Tyrnaviae

Lumnitzer S (1791) Flora Posoniensis exhibens plantas circa Posonium sponte crescents secundum systema sexuale Linneanum digestas. Lipsiae

Endlicher S (1830) Flora Posoniensis, exhibes plantas circa Posonium sponte crescentes aut frequentius cultas, methodo naturali dispositas Bratislava (Posonii)

Letz R (2000) Flora Bratislavy po dvock storociach od vyjdenia Lumnitzerovho diela Flora Posoniensis (The Flora of Braislava two centuries after publication of Lumnitzer's work *Flora Posoniensis*). Bull Slov Bot Spolcn Bratislava 22: 235–246

Spain

Madrid

Lopez Gonzalez GA, Lopez Jimenez N (1991) Flora de Madrid (Flora of Madrid) Real Jardin Botanico de Madrid

Switzerland

Basle/Basel

Bauhin C (1622) Catalogus plantarum circa Basileam sponte nascentium (A catalogue of plants growing around Basel) Basel

Brodtbeck T, Zemp M, Frei M, Kienzle M, Knecht D (1997) Flora von Basel und Umgebung 1980-1996, Teil I (Flora of Basle and vicinity, vol. 1). Mitteilungen Naturf Ges beider Basel 2: 1–543

Brodtbeck T, Zemp M, Frei M, Kienzle M, Knecht D (1998) Flora von Basel und Umgebung 1980-1996, Teil II (Flora of Basle and vicinity vol. 2) Mitteilungen Naturf Ges beider Basel 3: 544–1003

Fribourg

Purro C, Kozlowski G (eds) (2002) Flore de la ville de Fribourg (Flora of Fribourg). Editions Iniversitaires, Fribourg

Genève (*Geneva*)

Reuter G F (1832) Catalogue détaillé des plantes vasculaires qui croissant naturellement aux environs de Genève (Catalogue Dynamic of the Vascular Plants growing naturally in the vicinity of Geneva). Cherbuliez, Genève

Weber C (1966) Catalogue dynamique de la flore de Genève (Catalogue Dynamic of the flora of Geneva) Boissiera 12: 1–259

Theurillat J-P, Schneider C, Latour C (2011) Atlas de la flore de Genève. Catalogue analytique et distribution de la flore spontané (Atlas of the flora of Geneva. Catalogue analytic and distribution of the spontaneous flora). Conservatoire and Jardin Botaniques, Ville de Geneve

Lausanne

Hoffer-Massard F, Bornand C, Vust M (2006) Flore de Lausanne et de sa region: 1) A la découverte de la nature en ville. Rossolis, Lausanne (Flora of Lausanne and the region: 1) Discovery of nature in the City

Droz J, Hoffer-Massard F, Vust M, Bornand C (2006) Flore de Lausanne et de sa region. 2) Composition de la flore et repartition des espèces (Flora of Lausanne and the region 2) composition of the Flora and distribution of the species). Rossolis Lausanne

Lucerne

Brun-Hool J (1994) Wildpflanzen in Luzern (Wild plants in Lucerne). Mitt Natf Ges Luzern 37: 287

Zurich

Landolt E (1992) Veranderungen der Flora de stadt Zurichin den letzen 150 jahr. (Changes in the flora of Zurich in the last 150 years). Bauhinia 10: 149–164

Landolt E (1993-1999) Beitrage zur Flora der Stadt Zurich I (Contributions to the flora of the city Zurich I) Bot Helv 104-109

Landolt E (2001) Flora der Stadt Zürich: 1984 – 1998 mit Zeichn. von Rosmarie Hirzel (Flora of Zurich: 1984–1998 with drawings from Rosmarie Hirzel) Basel, Birkhäuser

The Netherlands

Amsterdam

Rombouts JHG, Merkus Doornik JJFHT (1852) Flora Amstelaedamensis (Flora of Amsterdam) Amsterdam

Prins J, Gorter A (1941) Planten van Amsterdam (Plants of Amsterdam). Nederlandsche Natuurhistorische Vereneniging afd Amsterdam

Bolman J (1976) Wilde Planten in en bij Amsterdam (Wild Plants in and around Amsterdam). Thieme, Zutphen

Denters T, Ruesink R, Vreeken B (1994) Van Muurbloem tot Straatmadelief; Wilde Planten in en rond Amsterdam (From Wall Street to Daisy Flower; Wild Plants in and around Amsterdam), Utrecht

Denters T, Vreeken B (1998) Flora-atlas van de regio Amsterdam (Flora Atlas of the Amsterdam region), Haarlem

Delft

an der Ham R, Sosef M (1987) Flora van Delft en Ormstreken (Flora of Delft and Surrounding Areas). Natuur- en Milieu-centrum De Papaver, Delft

Maastricht

Graatsma BG (2003) De flora van de omstreken van Maastricht in de 19 eeuw (The environs of Maastricht flora in the 19th century). Stichting Natuurpublicaties Limburg, Maastricht

Rotterdam

Balke NPW (1994) Vegetatie op het Rotterdamse puin (Vegetation on the Rotterdam rubble). Weer en Wind 8: 33–37

Andeweg R, Florusse P (2002) Vreemde planten in Rotterdam, Stadt ecologische reeks (Foreign plants in Rotterdam) Ecological series NR4 Natuurmuseum Rotterdam. 128 pp

Wales

Cardiff

Storrie J (1886) The Flora of Cardiff—a Descriptive List. Cardiff Naturalists' Society, Cardiff

Chepstow

Shoolbred WA (1920) The Flora of Chepstow. London

Swansea

Dillwyn LW (1848) Materials for a Fauna and Flora of Swansea and the neighbourhood

Chapter 9
Protozoans and Viruses

Protozoans As far as it can be ascertained there are no "comprehensive" or relatively comprehensive accounts of the protozoans of any city in Europe. It is a grave omission.

Viruses As far as it can be ascertained there are no "comprehensive" or relatively comprehensive accounts of the viruses of any city in Europe. It is a grave omission.

© Springer International Publishing Switzerland 2016

J.G. Kelcey, *Provisional Bibliography of Atlases, Floras and Faunas of European Cities: 1600–2014*, DOI 10.1007/978-3-319-31120-3_9

Annex I

Individuals and Organisations Approached for References

Key	
B	Birds
G	General
IN	Invertebrates
O	Other information
P	Plants
V	Vertebrates (other)

For the sake of completeness the compiler has also listed those countries for which he could not find contacts, despite considerable efforts. The museums were identified from Saur KG (2001) Museums of the World. 8th ed.

I have made my best endeavours to provide the information as to the affiliations of those who are listed below. I am sorry for the omissions, whatever the cause (in some cases the failure of my memory or filing system) and to those who are offended, my sincere apologies; it is not for the lack of effort: days have been spent searching books, documents and files but to no avail, eventually I gave up in exasperation.

Europe

British Phycological Society (Jane Pottas) (P)
Britsish Mycological Society (Norman Possett) (O)
CONTUREC—COmpetence NeTwork URban Ecology (O)
European Commission (Directorate-General, Environment) (G)
Hanel L Prof. RNDr. CSc. (V)
Harabiš F Dr. (V)
International Union for the Conservation of Nature (G)

© Springer International Publishing Switzerland 2016

J.G. Kelcey, *Provisional Bibliography of Atlases, Floras and Faunas of European Cities: 1600 –2014*, DOI 10.1007/978-3-319-31120-3

Ion M (IN)
Manu M Dr. (IN)
Muller N Prof. Dr. (G)
Sukopp H Prof. Dr. (G)
Kadlec T (IN)
University of Cranfield (Dr S. Hallett) (O)
Urban Nature Research Group—Toth-Ronkay M Dr. (G)
URBIO—International Network Urban Biodiversity and Design (G)
Werner P (G)

Albania

Natural History Museum, Tirane (G)

Andorra

No contacts

Armenia

No contacts

Austria

Holzer T Dr. (B)
Mrkvicka A Dr. (P)
Sziemer P Dr. (B)

Azerbaijan

No contacts

Belarus

Natural History Museum. Belovezskaja Pusca (G)

Belgium

Godefroid S Dr. (P)
Jacob J-P (B)
Weiserbs A (B)

Bosnia and Herzegovina

No contacts

Bulgaria

Dimitrov D Assoc. Prof. Dr. (P)
Geogiev D Dr. (V)
Iankov P Dr. (B)
Ivanov I (P)
Mollov I Assist. Prof. (V)
Nedyalkov N (V)

Popov V Assoc. Prof. Dr. (V)
Spassov N Assoc Prof. Dr. (V)
Stoyanov A (P)
Stoyneva M. Prof. (P)
Stefanov T (V)
Tzankov N Dr. (V)

Croatia

Natural History Museum, Brijuni (G)
Jasprica N (P)
Natural History Museum, Split (G)

Cyprus

Natural History Museum, Guzelyurt (G)

Czech Republic

Bejcek V (B)
Pysek P Dr. (P)
Sadlo J Dr. (P)
Sprynar P (P)
St'astny K Dr. (G)

Denmark

Natural History Museum, Lintrup (G)

Eire

No contacts

England

Abingdon Naturalists Society (Michael Bloom) (G)
Altrincham and District Natural History Society (G)
Banbury Natural History Society (G)
Barnsley Natural History Society (G)
Basingstoke Natural History Society (G)
Bath Natural History Society (Rob Randall) (G)
Birmingham Natural History Society (G)
Bishop's Stortford Natural History Society (G)
Botanical Society of the British Isles (P)
Bournemouth Natural Science Society (R. J. Chapman) (G)
Bristol Mammal Group (V)
Bristol Naturalists' Society (G)
Bristol Ornithological Club (B)
Bristol Regional Environmental Records Centre (Abigail Pedlow) (G)
Cambridge Natural History Society (G)
Carlisle Natural History Society (G)

Chesham and District Natural History Society (Andrea Polden) (G)
Chorley and District Natural History Society (Phil Kirk) (G)
Cornwall Environmental Records Centre (G)
Crawley, M. Prof. (P)
Croydon Natural History Society (G)
Derby Natural History Society (G)
Harrogate and District Naturalists' Society (Val Smith) (G)
Harrow Natural History Society (G)
Hillingdon Natural History Society (Ann Rix) (G)
Hull Natural History Society (Richard Middleton) (G)
Lancashire and Cheshire Fauna Society (G)
Leeds Philosophical and Literary Society (North ACT Prof. and Seaward M Dr.) (G)
Leicester Philosophical and Natural History Society (G)
Leigh Ornithological Society (B)
Liverpool Museum (G)
Lowestoft and North Suffolk Field Naturalist Club (G)
London Natural History Society (Edgington Dr. J) (G)
Loughborough Natural History Society (G)
Lymington Natural History Society (G)
Mammal Society (V)
Manchester Microscopical and Natural History Society (G)
Market Harborough Natural History Society (G)
Merseyside Naturalists' Association (Barbara Lee) (G)
Mitchell-Jones, Tony Dr. (V)
Natural History Museum, London (formerly the British Museum (Natural History)) (Lucy Carter) (G)
Newbury and District Field Club (G)
Oldham and District Natural History Society (Patricia Francis) (G)
Oxford (Hilary Killick) (G)
Preston Bird Watching and Natural History Society (G)
Reading and District Natural History Society (Malcolm Storey) (G)
Reading Ornithological Club (B)
Rochdale Field Natural History Society (G)
Royal Botanic Gardens, Kew (Craig Brough) (B)
Royal Entomological Society (IN)
Ruislip Natural History Society (G)
Self, A (B)
Sivell D (IN)
Sorby Natural History Society (G)
Stace, CA Prof (O)
University of Bristol; (Harris S Prof.) (V)
University of Exeter (V)
University of Oxford (McDonald D Dr.) (V)
Warwick Natural History Society (G)
Wells Natural History Society (G)

Worthing Natural History Society (G)
York Museum Trust (Rachel Wode) (V)
Note: The details of 46 "urban" Natural History Societies listed in a Directory of British Natural History Societies were not found.

Estonia

Uustal M (G)

Finland

Niemela J Prof. (IN)
Tianen J (B)
Vilisics F (IN)
Vuoisalo T (P)

France

Bertoncini A (P)
Clergeau P Prof. (V and IN)
Daniel H (P)
Machon N (P)
de Massary J-C (V and IN)
Muratet A (P)
Pellissier V (P)

Georgia

Natural History Museum, Batumi (G)
Natural History Museum, Chinvali (G)

Germany

Bendzko U (G)
Klausnitzer B Prof. Dr. (IN)
Malten A (G)
Mulsow R Dr. (B)
Rheinwald G (B)
Westphal U Dr. (G)
Witt K Dr. (B)
Zizka G Prof. Dr. (G)

Greece

Christodoulakis D (P)
Handlidu E (P)
Kokkini S (P)

Hungary

Hadarics T Dr. (G)
Gabor P (G)

Vilisics F (G)
Kiraly G (G)
Toth-Ronkay M Dr. (G)
Tibor M (IN)
Torok J (IN)
Tothmeresz B Prof. (IV)

Iceland

Natural History Museum, Reykjavik (G)

Italy

Celesti-Grapow L (P)
Cigini B (B)
Dinetti M (B)
Fraissenet M (B)
Iberite M Dr. (P)
Zapparoli M Prof. (V and IN)

Kosovo

No contacts

Latvia

No contacts

Liechtenstein

No contacts

Lithuania

No contacts

Luxembourg

Natural History Museum, Luxembourg (G)

Macedonia

No contacts

Malta

Natural History Museum, Victoria (G)

Moldova

No contacts

Monaco

No contacts

Montenegro

No contacts

Northern Ireland

Belfast Naturalists' Field Club (G)
McCaughey W (G)

Norway

Natural History Museum, Oslo (G)

Poland

Biadun W Dr. (B)
Galera H Dr. (P)
Jackowiak B Prof. Dr. (P)
Luniak M Prof. Dr. (B and O)
Sudnik-Wojcikowska B Prof. Dr. (P)

Portugal

Costa H (B)
Geraldes P (B)
Natural History Museum, Lisboa (G)

Romania

Onete M Dr. (G)
Pauca-Comanescu M Dr. (P)

Russia Federation

Ignetieva M Prof. Dr. (P)
Khrabryi V (B)
Kikolay (P)
Konechnaya G (P)
Konstantinov V Prof. (B)
Llminskikh G (P)
Shvetsov A Ass. Prof. Dr. (P)
Zakharov R (B)

San Marino

No contacts

Scotland

Crawley M Prof. (P)
Edinburgh Natural History Society (G)
Glasgow Natural History Society (Richard Weddle) (G)
Geoff Hancock, The Hunterian (Zoological Museum), University of Glasgow (G)

Serbia

Natural History Museum, Beograde (G)
Tucakov M (B)

Slovakia

Ferakova V Assoc. Prof. RNDr. CSc. (P)
Feriancova-Masarova, Z Prof. Dr. (B)
Jarolimek I RNDr. CSc. (P)
Kalivodova E Dr. (B)

Slovenia

No contacts

Spain

Dana E (P)
Guerrero M (P)
Garcia-de-Lomas J (P)
Murgui E (B)

Sweden

Hedblom M Dr. (G)
Ignetieva M Prof. Dr. (P)
Natural History Museum, Gotëborg (G)
Natural History Museum, Stockholm (G)

Switzerland

Baumann N (B)
Conservatoire et jardin botanique, Genève (Loizeau P-A) (G)
Geneve, Directore Generale de la nature et du paysage. (G).
Landolt E Prof. Dr. (P)
Museum d'histoire Naturell et Musee d'histoire des Sciences, Geneve (Danielle Dacrouez) (G)
Service des Espaces Verts et de L'Environment, Geneve (Daniel Oerti) (G)

The Netherlands

Denters T (P)
Vreekan B (P)
Weeda E Dr. (P)
de Zwarte N Dr. (V and IN)

Ukraine

Chmelnickij, Natural History Museum (G)
Ivano-Franovsk Natural History Museum (G)
Kirovgrad, Natural History Museum (G)

Lviv Natural History Museum (G)
Uzgorod Natural History Museum (G)
Vinnica Natural History Museum (G)

Vatican City

No contacts

Wales

Cardiff Natural History Society (G)
Countryside Council for Wales (G)
National Museum of Wales (G)

Annex II

Number of Animals (Invertebrates) References for Each Country

Key			
1	Acari	9	Dermaptera
2	Annelida	10	Dictyoptera
3	Anoplura	11	Diplopoda (see Myriapoda)
4	Arachnida	12	Diptera
5	Chilopoda (see Myriapoda)	13	Ephemeroptera
6	Coleoptera	14	Hemiptera/Heteroptera/Homoptera
7	Collembola	15	Hymenoptera
8.	Decapoda		

Country	1	2	3	4	5	6	7	8	9	10	11	12	13	14	15
Albania	0	0	0	0	0	0	0	0	0	0	0	0	0	0	0
Andorra	0	0	0	0	0	0	0	0	0	0	0	0	0	0	0
Armenia	0	0	0	0	0	0	0	0	0	0	0	0	0	0	0
Austria	0	0	0	0	0	0	0	0	0	0	0	0	0	0	0
Azerbaijan	0	0	0	0	0	0	0	0	0	0	0	0	0	0	0
Belarus	0	0	0	0	0	0	0	0	0	0	0	0	0	0	0
Belgium	0	0	0	0	0	0	0	0	0	0	0	0	0	0	0
Bosnia and Herzegovina	0	0	0	0	0	0	0	0	0	0	0	0	0	0	0
Bulgaria	0	0	0	0	0	1	0	0	0	0	0	0	0	1	1
Croatia	0	0	0	0	0	0	0	0	0	0	0	0	0	0	0
Cyprus	0	0	0	0	0	0	0	0	0	0	0	0	0	0	0
Czech Republic	0	0	0	1	0	1	0	0	0	0	0	0	0	0	0
Denmark	0	0	0	0	0	0	0	0	0	0	0	0	0	0	0
Eire	0	0	0	0	0	0	0	0	0	0	0	0	0	0	0

© Springer International Publishing Switzerland 2016
J.G. Kelcey, *Provisional Bibliography of Atlases, Floras and Faunas of European Cities: 1600 –2014*, DOI 10.1007/978-3-319-31120-3

Country	1	2	3	4	5	6	7	8	9	10	11	12	13	14	15
England	0	0	1	6	0	2	0	0	0	0	0	8	0	19	0
Estonia	0	0	0	0	0	0	0	0	0	0	0	0	0	0	0
Finland	0	0	0	0	0	0	0	0	0	0	0	0	0	0	1
France	0	0	0	0	0	0	0	0	0	0	0	0	0	0	0
Georgia	0	0	0	0	0	0	0	0	0	0	0	0	0	0	0
Germany	0	0	0	0	0	0	0	0	0	0	0	0	0	0	0
Greece	0	0	0	0	0	0	0	0	0	0	0	0	0	0	0
Hungary	0	0	0	0	0	0	0	0	0	0	0	0	0	0	0
Iceland	0	0	0	0	0	0	0	0	0	0	0	0	0	0	0
Italy	0	0	0	0	0	0	0	0	0	0	0	0	0	0	0
Kosovo	0	0	0	0	0	0	0	0	0	0	0	0	0	0	0
Latvia	0	0	0	0	0	0	0	0	0	0	0	0	0	0	0
Liechtenstein	0	0	0	0	0	0	0	0	0	0	0	0	0	0	0
Lithuania	0	0	0	0	0	0	0	0	0	0	0	0	0	0	0
Luxembourg	0	0	0	0	0	0	0	0	0	0	0	0	0	0	0
Macedonia	0	0	0	0	0	0	0	0	0	0	0	0	0	0	0
Malta	0	0	0	0	0	0	0	0	0	0	0	0	0	0	0
Moldova	0	0	0	0	0	0	0	0	0	0	0	0	0	0	0
Monaco	0	0	0	0	0	0	0	0	0	0	0	0	0	0	0
Montenegro	0	0	0	0	0	0	0	0	0	0	0	0	0	0	0
Northern Ireland	0	0	0	0	0	0	0	0	0	0	0	0	0	0	0
Norway	0	0	0	0	0	0	0	0	0	0	0	0	0	0	0
Poland	1	0	0	2	0	7	1	0	0	0	0	13	0	1	7
Portugal	0	0	0	0	0	0	0	0	0	0	0	0	0	0	0
Romania	0	0	0	0	0	0	0	0	0	0	0	0	0	0	0
Russian Federation	0	0	0	0	0	0	0	0	0	0	0	0	0	0	0
San Marino	0	0	0	0	0	0	0	0	0	0	0	0	0	0	0
Scotland	0	0	0	0	0	0	0	0	0	0	0	0	0	0	0
Serbia	0	0	0	0	0	0	0	0	0	0	0	0	0	0	0
Slovakia	0	0	0	0	0	1	0	0	0	0	0	0	0	0	0
Slovenia	0	0	0	0	0	0	0	0	0	0	0	0	0	0	0
Spain	0	0	0	0	0	0	0	0	0	0	0	0	0	0	0
Sweden	0	0	0	0	0	0	0	0	0	0	0	0	0	0	0
Switzerland	0	0	0	0	0	0	0	0	0	0	0	0	0	0	0
The Netherlands	0	0	0	0	0	0	0	0	0	0	0	0	0	0	0
Ukraine	0	0	0	0	0	0	0	0	0	0	0	0	0	0	0
Vatican City	0	0	0	0	0	0	0	0	0	0	0	0	0	0	0
Wales	0	0	0	0	0	0	0	0	0	0	0	0	0	0	0
Total 52	1	0	1	9	0	12	1	0	0	0	0	21	0	21	9

Key			
16	Isopoda	23	Odonata
17	Isoptera	24	Oligochaeta
18	Lepidoptera	25	Orthoptera
19	Mollusca	26	Phasmida
20	Myriapoda	27	Platyhelminthes
21	Myriapoda	28	Plecoptera
22	Neuroptera	29	Siphonoptera

Country	16	17	18	19	20	21	22	23	24	25	26	27	28	29
Albania	0	0	0	0	0	0	0	0	0	0	0	0	0	0
Andorra	0	0	0	0	0	0	0	0	0	0	0	0	0	0
Armenia	0	0	0	0	0	0	0	0	0	0	0	0	0	0
Austria	0	0	0	0	0	0	0	0	0	0	0	0	0	0
Azerbaijan	0	0	0	0	0	0	0	0	0	0	0	0	0	0
Belarus	0	0	0	0	0	0	0	0	0	0	0	0	0	0
Belgium	0	0	0	0	0	0	0	0	0	0	0	0	0	0
Bosnia and Herzegovina	0	0	0	0	0	0	0	0	0	0	0	0	0	0
Bulgaria	0	0	4	0	1	0	0	0	0	0	0	0	0	0
Croatia	0	0	0	0	0	0	0	0	0	0	0	0	0	0
Cyprus	0	0	0	0	0	0	0	0	0	0	0	0	0	0
Czech Republic	0	0	0	2	0	0	0	0	0	0	0	0	0	0
Denmark	0	0	0	0	0	0	0	0	0	0	0	0	0	0
Eire	0	0	0	0	0	0	0	0	0	0	0	0	0	0
England	0	0	60	2	0	0	1	4	0	3	0	0	0	1
Estonia	0	0	0	0	0	0	0	0	0	0	0	0	0	0
Finland	0	0	0	0	0	0	0	0	0	0	0	0	0	0
France	0	0	0	0	0	0	0	0	0	0	0	0	0	0
Georgia	0	0	0	0	0	0	0	0	0	0	0	0	0	0
Germany	0	0	0	0	0	0	0	3	0	2	0	0	0	0
Greece	0	0	0	0	0	0	0	0	0	0	0	0	0	0
Hungary	0	0	0	0	0	0	0	0	0	0	0	0	0	0
Iceland	0	0	0	0	0	0	0	0	0	0	0	0	0	0
Italy	0	0	0	0	0	0	0	0	0	0	0	0	0	0
Kosovo	0	0	0	0	0	0	0	0	0	0	0	0	0	0
Latvia	0	0	0	0	0	0	0	0	0	0	0	0	0	0
Liechtenstein	0	0	0	0	0	0	0	0	0	0	0	0	0	0
Lithuania	0	0	0	0	0	0	0	0	0	0	0	0	0	0
Luxembourg	0	0	0	0	0	0	0	0	0	0	0	0	0	0
Macedonia	0	0	0	0	0	0	0	0	0	0	0	0	0	0
Malta	0	0	0	0	0	0	0	0	0	0	0	0	0	0
Moldova	0	0	0	0	0	0	0	0	0	0	0	0	0	0

Country	16	17	18	19	20	21	22	23	24	25	26	27	28	29
Monaco	0	0	0	0	0	0	0	0	0	0	0	0	0	0
Montenegro	0	0	0	0	0	0	0	0	0	0	0	0	0	0
Northern Ireland	0	0	0	0	0	0	0	0	0	0	0	0	0	0
Norway	0	0	0	0	0	0	0	0	0	0	0	0	0	0
Poland	1	0	1	0	1	0	1	0	3	0	0	0	0	0
Portugal	0	0	0	0	0	0	0	0	0	0	0	0	0	0
Romania	0	0	1	0	0	0	0	0	0	0	0	0	0	0
Russian Federation	0	0	0	1	0	0	0	0	0	0	0	0	0	0
San Marino	0	0	0	0	0	0	0	0	0	0	0	0	0	0
Scotland	0	0	0	0	0	0	0	0	0	0	0	0	0	0
Serbia	0	0	0	0	0	0	0	0	0	0	0	0	0	0
Slovakia	0	0	0	0	0	0	0	1	0	0	0	0	0	0
Slovenia	0	0	0	0	0	0	0	0	0	0	0	0	0	0
Spain	0	0	0	0	0	0	0	0	0	0	0	0	0	0
Sweden	0	0	0	0	0	0	0	0	0	0	0	0	0	0
Switzerland	0	0	0	0	0	0	0	0	0	0	0	0	0	0
The Netherlands	0	0	0	0	0	0	0	0	0	0	0	0	0	0
Ukraine	0	0	0	0	0	0	0	0	0	0	0	0	0	0
Vatican City	0	0	0	0	0	0	0	0	0	0	0	0	0	0
Wales	0	0	0	0	0	0	0	0	0	0	0	0	0	0
Total 52	1	0	66	5	2	0	2	8	3	5	0	0	0	1

Grand Total Invertebrate references = 168

Annex III

Number of Animals (Vertebrates) References for Each Country

Country	Amphibians/reptiles	Birds	Fish	Mammals
Albania	0	0	0	0
Andorra	0	0	0	0
Armenia	0	0	0	0
Austria	0	3	0	0
Azerbaijan	0	0	0	0
Belarus	0	0	0	0
Belgium	0	7	0	0
Bosnia and Herzegovina	0	0	0	0
Bulgaria	1	1	0	0
Croatia	0	0	0	0
Cyprus	0	0	0	0
Czech Republic	2	8	0	3
Denmark	0	0	0	0
Eire	0	0	0	0
England	10	70	6	28
Estonia	0	0	0	0
Finland	0	0	0	0
France	1	4	0	0
Georgia	0	0	0	0
Germany	25	28	0	0
Greece	0	0	0	0
Hungary	0	1	0	0
Iceland	0	0	0	0
Italy	2	52	0	0

© Springer International Publishing Switzerland 2016

J.G. Kelcey, *Provisional Bibliography of Atlases, Floras and Faunas of European Cities: 1600–2014*, DOI 10.1007/978-3-319-31120-3

Country	Amphibians/reptiles	Birds	Fish	Mammals
Kosovo	0	0	0	0
Latvia	0	0	0	0
Liechtenstein	0	0	0	0
Lithuania	0	2	0	0
Luxembourg	0	0	0	0
Macedonia	0	0	0	0
Malta	0	0	0	0
Moldova	0	0	0	0
Monaco	0	0	0	0
Montenegro	0	0	0	0
Northern Ireland	0	0	0	0
Norway	0	0	0	0
Poland	1	20	0	0
Portugal	0	2	0	0
Romania	0	0	0	0
Russian Federation	0	9	0	0
San Marino	0	0	0	0
Scotland	1	0	0	1
Serbia	0	1	0	0
Slovakia	0	0	0	0
Slovenia	0	0	0	0
Spain	0	1	0	0
Sweden	0	0	0	0
Switzerland	1	1	0	0
The Netherlands	0	6	0	0
Ukraine	0	2	0	0
Vatican City	0	0	0	0
Wales	0	0	0	0
Total 52	44	214	6	32

Grand Total Animals (Vertebrate) references = 302

Annex IV

Number of Non-vascular and Vascular Plants and Fungi References for Each Country

Country	Plants (non-vascular)	Plants (vascular)	Fungi	Fungi (lichenised)
Albania	0	0	0	0
Andorra	0	0	0	0
Armenia	0	0	0	0
Austria	4	8	0	2
Azerbaijan	0	0	0	0
Belarus	0	0	0	0
Belgium	1	3	2	2
Bosnia and Herzegovina	0	0	0	0
Bulgaria	0	1	0	0
Croatia	0	1	0	0
Cyprus	0	0	0	0
Czech Republic	0	3	0	0
Denmark	0	0	0	0
Eire	0	2	0	0
England	7	109	1	4
Estonia	0	0	0	0
Finland	0	1	0	0
France	0	1	0	0
Georgia	0	0	0	0
Germany	0	36	0	0
Greece	0	4	0	0
Hungary	0	0	0	0
Iceland	0	0	0	0

© Springer International Publishing Switzerland 2016

J.G. Kelcey, *Provisional Bibliography of Atlases, Floras and Faunas of European Cities: 1600 –2014*, DOI 10.1007/978-3-319-31120-3

Country	Plants (non-vascular)	Plants (vascular)	Fungi	Fungi (lichenised)
Italy	1	17	0	1
Kosovo	0	0	0	0
Latvia	0	0	0	0
Liechtenstein	0	0	0	0
Lithuania	0	0	0	0
Luxembourg	0	0	0	0
Macedonia	0	0	0	0
Malta	0	0	0	0
Moldova	0	0	0	0
Monaco	0	0	0	0
Montenegro	0	0	0	0
Northern Ireland	0	2	0	0
Norway	0	0	0	0
Poland	0	18	1	0
Portugal	0	0	0	0
Romania	0	0	0	0
Russian Federation	0	11	0	1
San Marino	0	0	0	0
Scotland	0	5	0	0
Serbia	0	0	0	0
Slovakia	1	4	0	0
Slovenia	0	0	0	0
Spain	0	1	0	0
Sweden	0	0	0	0
Switzerland	0	13	0	0
The Netherlands	0	9	0	0
Ukraine	0	0	0	0
Vatican City	0	0	0	0
Wales	0	3	0	0
Total 52	14	253	4	10

Grand Total Plants (Vascular and Non-Vascular) references=281

Annex V

Examples of General References

This Annex contains examples of general accounts of the natural history, flora and fauna of cities that are not included in the main bibliography. For that reason it is important to emphasis that they are examples and that there is scope for a bibliography on the natural history of cities *sensu lato*.

Europe

John G Kelcey, Goetz Rheinwald (2005) Birds in European Cities. Ginster Verlag, St. Katharinen

John G Kelcey, Norbert Muller (2011) Plants and Habitats of European Cities. Springer, New York

John G Kelcey (2015) Vertebrates and Invertebrates of European Cities: Selected Non-Avian Fauna. Springer, New York.

England

Liverpool

Byerley I (1854) The Fauna of Liverpool. Appendix to the Proc. Lit. & Phil. Soc. Liverpool 1853-1854

Byreley I (1856) The Fauna of Liverpool

London

Buckell F S, Prout L B (1898-1902) The fauna of the London District: Lepidoptera. Transactions, City of London Entomological and Natural History Society 8: 51–63; 9: 66–80; 10: 62–74 and 11: 54–58

Fitter RSR (1945) London's Natural History. Collins, London

Oldham (now in Greater Manchester)

Kidd L (1977) Oldham's Natural History

© Springer International Publishing Switzerland 2016

J.G. Kelcey, *Provisional Bibliography of Atlases, Floras and Faunas of European Cities: 1600 –2014*, DOI 10.1007/978-3-319-31120-3

Italy

Rome

Zapparoli M (ed) (1997) Gli Insetti di Roma. (Insects of Rome) Fratelli Palombi Editori Roma, Comunedi Roma, Assessorato, Ambiente

Taglianti AV, Zapparoli M (2006) Insetti di Roma; biodiversita in un ecosistema urban (Insects of Rome; biodiversity in an urban environment). Museo di Zoologia, Universita Degli Studi 'La Sapienza,' Edizioni Belvedere

Russia

Moscow

Formozovan AN (1947) Nature of Moscow and the Moscow Region. Izd Acad Nauk SSSR Moskwa: 287–370

Scotland

Glasgow

Scott-Elliot GF, Laurie M, Murdoch JB (1901) Fauna, flora and geology of the Clyde area Glasgow. Local Committee for the meeting of the British Association

Index

A
Aberdeen, 108
Acari, 2, 24, 29
Aldermaston, 39
Alexandroupolis, 102
Algae, 2, 28, 85
Alkmaar, 74
Altdorf, 99
Altenburg, 99
Altötting, 52
Amphibia/amphibians, 14, 26–27, 78
Amsterdam, 74, 110
Ancona, 103
Andover, 92
Annelida, 2, 24
Anoplura, 24, 29
Ants, 2, 25, 37–38
Aosta, 103
Arachnida, 2, 24, 30
Armenia, 6
Aschaffenburg, 47
Ashton-under-Lyne (now in Greater Manchester), 92
Augsburg, 99
Austria, 56, 82, 86, 90–91
Aves (Birds), 27, 56
Azerbaijan, 6

B
Bacteria and related organisms, 28, 79
Bad Aussee, 90
Baden-Württemberg, 53
Banbury, 58
Banska Bystrica, 32
Basildon, 92
Basle/Basel, 73, 109
Bath, 92
Bayreuth, 53
Bees, 2, 25, 37–38
Beetles, 2, 24, 31–32
Belfast, 105
Belgium, 7, 57, 81–83, 86, 91
Beograd (Belgrade), 7, 73
Bergamo, 65
Berkhamstead, 39
Berlin, 12, 27, 47, 53, 62, 99–100
Bern (Berne), 56
Biddulph, 92
Bielefeld, 63
Biella, 65
Birds, 3, 14, 27, 56–58
Birmingham, 4, 58, 83, 92
Bishops Stortford, 40, 93
Blankenburg, 53
Bologna, 55, 65
Bolton, 93
Bonn, 63
Bournemouth, 93
Brabant, 57
Bracknell, 40
Bradford, 93
Bratislava, 87, 108–109
Braunschweig, 100
Breda, 74
Brentwood, 40
Brescia, 65
Brighton, 58

© Springer International Publishing Switzerland 2016

J.G. Kelcey, *Provisional Bibliography of Atlases, Floras and Faunas of European Cities: 1600–2014*, DOI 10.1007/978-3-319-31120-3

Bristol, 40, 47, 58, 75, 93
Brno, 51, 75
Bruxelles (Brussels), 83, 86, 91
Bryophytes, 28, 86–87, 92
Bucuresti (Bucharest), 43
Bugs, 25, 35–37
Bulgaria, 31, 35, 37, 39, 45, 51, 57, 91
Burton-on-Trent, 93
Bury (now in Greater Manchester), 58, 75
Butterflies, 25, 39–42

C
Cagliari, 66
Calderdale, 58
Calw, 53
Cambridge, 28, 93
Camerino, 103
Cardiff, 111
Centipedes, 24, 25, 31, 45
Chalfont, 40
Chelmsford, 47
Chemnitz, 63, 100
Chepstow, 111
Cheshunt, 40
Chester, 58
Chieti, 103
Chilopoda, 24, 45
Chorleywood, 58, 61
Christchurch, 93
Cirencester, 93
Cockroaches, 25
Cogoleto, 103
Colchester, 26
Coleoptera, 2, 24, 31–32
Collembola, 24, 33
Copenhagen, 25, 58
Cossato, 66
Cottbus, 53
Crema (CR), 66
Cremona, 66
Crewkerne, 59
Croatia, 91
Cyprus, 6
Czech Republic, 30, 32, 44, 51–52, 57–58, 75, 91–92
Częstochowa, 71

D
Danzig (Gdansk), 100, 105
Darmstadt, 100
Dartford, 59
Debrecen, 65
Decapoda, 24
Delft, 74, 110
Denmark, 58
Derby, 75, 94
Dermaptera, 24
Dessau, 101
Dewsbury, 94
Dictyoptera, 25
Diplopoda, 25, 33, 45
Diptera, 2, 25, 34, 49
Doncaster, 40, 94
Dortmund, 63
Douai, 62
Dragonflies and Damselflies, 26, 46
Dublin, 92
Dubrovnik, 47
Duisburg, 101
Dusseldorf, 53, 63–65

E
Earthworms, 48
Earwigs, 24
Eastbourne, 59
Edinburgh, 56, 59, 108
Eindhoven, 74
Eire, 92
Eisenerz, 90
Elberfeld, 53
Emden, 63
England, 4, 12, 20, 29–36, 39–44, 46, 47, 49, 52, 58–62, 75–78, 82, 83, 87, 92–98
Ephemeroptera, 25
Essen, 64, 65, 101
Euphyllophytes, 13
Euskirchen, 101

F
Ferns, 28, 91, 95, 96, 98, 100, 101
Finland, 37, 98
Firenze (Florence), 66, 103
Fish, 13, 14, 27, 77–78
Fleas, 2, 3, 26, 29, 49
Flies, 25, 33, 44
Forests/woodlands, 23
Forli, 66
Formby, 40, 59
France, 18, 28, 52, 62, 99

Frankfurt, 27, 101
Fribourg, 109
Fungi, 2, 3, 14, 21, 28, 81–83
Fürth, 54

G
Gardens, Parks and Industrial Habitats, 23
Gastropoda/Gastropods, 45
Gela (Sicilia), 87, 103
Gelsenkirchen, 54
Genève (Geneva), 109
Genova (Genoa), 67, 104
Geology, 4, 7, 16, 23, 77
Geomorphology, 7, 23
Georgia, 6
Germany, 7, 27, 47, 49, 52–55, 62–65, 99–102
Giessen, 17, 18, 49
Glasgow, 77, 108
Gliwice, 105
Glogowek, 105
Gorzow, 71
Göttingen, 101
Grasshoppers and crickets, 26, 48
Grasslands, 23
Greater Manchester, 11, 12, 40, 58, 60, 61, 87, 92, 93, 95–97
Greece, 6, 18, 102
Greenland, 6
Grosseto, 67

H
Halberstadt, 64
Halifax, 40, 52, 58, 75, 82, 94
Halle, 64
Hamburg, 54, 64, 101
Harrogate, 41
Hatfield, 41
Heathland, 23
Helsinki, 28, 37, 98
Hemiptera-Heteroptera, 25, 35, 36
Hepatophyta, 87
Hessen, 49
Heywood (now in Greater Manchester), 59
Hitchin, 35, 41
Hof, 54
Horazdovice, 91–92
Hradec Kralove, 44
Hull (correct name is Kingston upon Hill), 94
Hungary, 65
Hymenoptera, 2, 25, 37–38

I
Isopoda, 25, 38
Isoptera, 25
Israel, 6
Italy, 4, 19, 55, 65–70, 87, 103–104
Izevsk, 107

J
Jaslo, 71
Jaworznie, 105

K
Kazakhstan, 6
King's Lynn, 94
Kloch, 90
Kosovo, 6
Kozle, 105
Krakow (Crakow), 105
Krk, 91
Krumegg, 90

L
Lacewings, 26, 46
La Spezia, 67
Latium, 67
Lausanne, 110
Lazio, 67
Leeds, 59, 93, 94
Leek, 94
Legnica, 105
Leipzig, 101
Lepidoptera, 25, 39–43
Leszno, 71
Letchworth, 41, 59
Leyton (now in Greater London), 41
Lice, 24, 29, 49
Lichen, 82, 83
Lichenised Fungi, 28, 81–83
Lido di Ostia, 67
Liguria, 55, 67
Linz, 56
Lipetsk, 107
Lisboa (Lisbon), 72
Lithuania, 70
Liverpool, 41, 59, 87, 94–95

Liverworts, 28, 106
Livorno, 68
Lodz, 71, 82, 105
London, 9–11, 16, 23–33, 35–36, 41–42, 44, 46, 47, 49, 52, 59–60, 76, 77, 83, 87, 95
Lubań Śląski, 106
Lublin, 71, 106
Lucerne, 110
Luton, 95
Lutsk, 74
Luxembourg, 27
Lvov, 74
Lycophytes, 13

M
Maastricht, 111
Madrid, 109
Magdeburg, 54
Mainz, 64
Malta, 18
Mammalia (mammals), 14, 27, 75–77
Manchester (now in Greater Manchester), 60, 95
Marine, 23
Marseille, 52
Mayflies, 25
Melle, 64
Mesolongi, 102
Middleton (now in Greater Manchester), 96
Milano (Milan), 68, 104
Mildenhall, 96
Millipedes, 25, 45
Mites, 2, 24, 29
Mollusca (Molluscs), 25, 36, 44–45
Monilophytes, 13
Monti Lepini, 68
Moscow, 45, 107
Mosses, 28, 87, 106
Moths, 39–42
Muelheim, 64, 65
Mühldorf, 54
Munchen (Munich), 101
Myriapoda, 24, 25, 31, 33, 45
Mysidacea, 26

N
Napoli (Naples), 68, 104
The Netherlands, 74, 110–111
Neuroptera, 26, 44, 46
Newbury, 42, 60, 92, 96
Niuwegein, 74
Northern Ireland, 105
Nottingham, 96

O
Oberhausen, 65
Odonata, 26, 46, 47
Oldham (now in Greater Manchester), 96
Oligochaeta, 24, 26, 48
Olsztyn, 71
Opole, 105
Orthoptera, 2, 24–26, 48, 49
Ortona, 104
Osnabruck, 102
Oswestry, 96
Oundle, 96
Oxford, 24, 27, 43, 49, 60, 96

P
Padova, 68
Palermo, 68
Pardubice, 57
Paris, 24–26, 62
Patras, 102–103
Pavia, 68
Petrozavodsk, 107
Phasmida, 26
Piacenza, 69
Pices, 27, 77–78, 92
Pisa, 69
Pischelsdorf, 90
Plants (vascular), 28, 90–111
Platyhelminthes, 24, 26
Plecoptera, 26, 36
Plovdiv, 51
Plymouth, 61, 96
Plzeň (Pilsen), 92
Poland, 7, 20, 29, 31–38, 45, 46, 48, 55, 71–72, 82, 105–106
Polkowice, 106
Portugal, 72
Poznan, 71, 106
Praha (Prague), 30, 32, 44, 52, 57, 75, 92
Preston, 96
Protozoans, 13, 14, 21, 28, 113
Pteridophytes, 89–111

R
Razgrad, 39
Reading, 40, 61, 97

Regensburg, 54, 65, 102
Reggio Emilia, 69
Reigate, 97
Reptilia (Reptiles), 27, 78
Rickmansworth, 61
Rochdale (now in Greater Manchester), 97
Roma (Rome), 55, 69, 83, 104
Romania, 7, 43
Rotterdam, 16, 111
Rovinj, 91
Royston, 97
Rugby, 97
Russian Federation, 45, 72–73, 83, 107–108

S
Saarbrücken, 49, 65
Saarlouis, 102
Saddleworth, 97
Salzburg, 90
San Dona di Piave, Venezia (Venice), 69
Sandbach, 61
Saratov, 108
Scarborough, 43, 44, 76
Schwandorf, 54
Scordia (Sicilia), 104
Scotland, 56, 77, 108
Scrub, 24
Serbia, 7, 47, 73
Sheffield, 31, 32, 34, 36, 43, 49, 52, 61, 76–78, 96, 97
Shoreham, 58, 61, 62
Shrimps, 24, 26
Siphonoptera, 26
Sliven, 39
Slovakia, 32, 87, 108–109
Sofia, 31, 35, 37, 57, 91
Soil, 4, 7, 16, 23, 29, 45
Southport, 61, 87
Spain, 6, 73, 109
Spalding, 61
Spermatophytes, 13
Spessart, 55
Springtails, 24, 33
Staines, 61
Steyning, 58, 61, 62
Stick Insects, 26
Stoneflies, 26
Stoneyhurst (Greater Manchester), 62
St. Petersburgh (formerly Leningrad), 73, 107
Stupsk, 71
Stuttgart, 55, 102
Sulechowa, 72
Swansea, 111
Świebodzina, 72
Swilengrad (Svilengrad), 39
Switzerland, 27, 56, 73, 109–110

T
Termites, 25
Thessaloniki, 103
Ticks, 2, 24, 29
Tirnovo (Turnovo), 39
Todmorden, 62, 97
Torino (Turin), 69
Torquay, 97
Trento, 69, 104
Treviso, 70
Tunbridge Wells, 97
Turkey, 6

U
Udine, 70
Ukraine, 74
Ulyanovsk, 108
Umbria, 70
Uplands, 24

V
Valencia, 73
Valsesia, 70
Varese, 70
Vatican City, 6
Verano Cusio Ossola, 70
Vercelli, 104
Vilinius, 70
Viruses, 13, 14, 21, 28, 113
Viterbo, 70
Voronezh, 73, 108

W
Wakefield, 98
Wales, 111
Warrington, 98
Warszawa (Warsaw), 29, 32–38, 45, 46, 48, 55, 72, 106
Wasps, 2, 25, 37–38
Water, Moving (Lotic), 24
Water, Still (Lentic), 24

Watford, 43, 98
Weissenburg-Gunzenhausen, 55
Weston-super-Mare, 98
Wetlands, 24
Widnes, 98
Wien (Vienna), 56, 82, 86, 91
Wilmslow, 62
Winchester, 62
Witten, 55
Wolfsburg, 102
Woodlice, 25
Worcester, 62
Worms, flat, 2, 16, 24, 26
Worms, segmented, 26
Worthing, 58, 61, 62
Wuppertal, 55, 102

Y
York, 44, 77, 78, 94, 98

Z
Zurich, 110

The manufacturer's authorised representative in the EU is Springer Nature Customer Service Centre GmbH, Europaplatz 3, 69115 Heidelberg, Germany. If you have any concerns regarding our products, please contact ProductSafety@springernature.com

Printed and bound by CPI Group (UK) Ltd, Croydon, CR0 4YY
12/08/2026
02194544-0002